HUME PAPERS ON PUBLIC POLICY
Volume 8, No. 2

ENVIRONMENT AND REGULATION
Edited by Andrea Ross

THE DAVID HUME INSTITUTE

HUME PAPERS ON PUBLIC POLICY
Volume 8 No. 2

ENVIRONMENT AND REGULATION
Edited by Andrea Ross

EDINBURGH UNIVERSITY PRESS

Edinburgh University Press
22 George Square, Edinburgh.

Typeset in Times New Roman by
WestKey Limited, Falmouth, Cornwall
and printed and bound in Great Britain by
Page Bros Ltd, Norwich Norfolk

A CIP record of this book is available from the British Library

ISBN 0 7486 1473 7

Contents

Contributors

Valerie A. Dickie is a Research Associate in the Division of Economics, School of Management, Heriot-Watt University

Nick Hanley is a Professor of Environmental Economics, University of Glasgow

Douglas MacMillan is a senior lecturer in Agricultural Economics in the Department of Agriculture, University of Aberdeen

Kathryn V. Last in a lecturer in Law in the Department of Law, University of Dundee

Colin T. Reid is Professor of Environmental Law in the Department of Law, University of Dundee

Andrea Ross is a senior lecturer in Law in the Department of Law, University of Dundee

John W. Sawkins is a lecturer in Economics in the Division of Economics, School of Management, Heriot-Watt University

Foreword

At all levels of government, the problem of environmental regulation is receiving ever more urgent attention. Often regulatory efforts rely on legal enforcement but, equally, alternative more market oriented mechanisms are utilised. The set of papers in this issue of *Hume Papers on Public Policy* contains a detailed examination of the scope and usefulness of the range of available policies in this area, from both a theoretical and practical perspective. The applied areas discussed include habitat conservation, wildlife (red deer and wild salmon), outdoor recreation sites, multi-use forests, and water. Policy instruments discussed range from legal codes to economic incentives. Distinctions are also drawn between policy in England and Wales and that in Scotland, now that certain powers in this area are devolved to the new Scottish Parliament. The impact of the new devolved powers in Scotland is the subject of one of the papers, and most of the applied case studies are set in Scotland.

These papers will be of interest to lawyers, economists, environmentalists and, indeed, to all those with a policy interest in the environment. We are grateful to Andrea Ross as the editor of this important issue of *Hume Papers on Public Policy*. The David Hume Institute is delighted to be able to publish this collection of essays which contribute in such a timely way to our understanding of this essential aspect of commercial life. As always, it is necessary to clarify that the Institute itself holds no collective view on these policy matters. We do, however, feel that we can recommend this work as a useful and provocative contribution to our understanding of an area of significant commercial importance.

Brian G M Main
Director
The David Hume Institute

Introduction

Science has had a huge impact on public policy in recent years. Discoveries about the earth's fragility, the plight of other species, and our mutual co-dependence have combined with continually improving communications to bring new issues into the policy-making arena. Globalisation is one such new issue. Global markets, world wide live telecommunications, international crime rings and transboundary environmental hazards such as acid rain and climate change have meant that most areas of policy making are now under some international influence and policy makers around the world must work together towards common goals. Another issue worth mentioning is the widespread acceptance of some notion of sustainable development. The most widely cited definition of sustainable development comes from the Bruntland Commission[1] which reported in 1987: "development which meets the needs of the present without compromising the ability of future generations to meet their own needs." Essentially, sustainable development means dealing with economic policy, social policy and environmental policy together to create sustainable solutions now and in the future. The term has been criticised as vague, imprecise and meaningless. That said, it is because of this vagueness that sustainable development has received world-wide acceptance.

To meet these new demands public policy has to adapt. Governments and policy makers are reviewing their policy objectives, reconsidering and re-evaluating their styles of regulation and in some instances even re-examining the structures and institutions which develop and implement them.

One such change has been a recognition of the need to consider the longer term and take a more strategic approach to policy making. This has led to the publication of white papers like *Modernising Government*[2] and *A Better Quality of Life – A strategy for sustainable development in the UK*[3] which states "For the future, we need ways to achieve economic, social and environmental objectives at the same time, and consider the longer term implications of decisions … This Strategy is a catalyst for that change … The Government will use the Strategy as a framework to guide its policies. It will encourage others to do the same." (paras 1.10–1.11).

---

[1] World Commission on Environment and Development , *Our Common Future* (1987) p. 43.
[2] HMSO, 1999 Cm 4310.
[3] HMSO, 1999 Cm 4345.

A second change in approach is the notion that decisions should not be made in isolation. Calls for integration, assimilation, joined up government, and mainstreaming all deal with this issue. Increasingly, agencies are either being given new (and often conflicting) objectives or they are being given 'balancing obligations' which require them at least to have regard to other concerns and in some instances to consult with other agencies. For example, the Forestry Commission is no longer simply concerned with timber production and now also has duties in regard to conservation. Appraisal mechanisms are used to consider the influence or impact of a particular policy or project on the environment, older people, internal and external markets etc. There have been institutional changes to create more joint bodies, such as the North South Ministerial Council in Ireland established as part of the devolution settlement in Northern Ireland, to develop consultation and co-operation on certain matters including the environment, within the island of Ireland.

Thirdly, modern government must be more open, transparent, inclusive and responsive. More information than ever is made available by the public sector to members of the public in the form of reports, public advice lines, and internet web-sites. These measures are often supported by guidance on good practice, citizens' charters and in some instances legislation. Furthermore, there are moves to consult more widely, not so much with individual members of the public, but with representative groups such as interest groups and stakeholders. Institutional changes are also being introduced. Bringing government closer to the people through devolution is one such change. Other important developments are the changes to the voting systems at all levels of government in the UK to allow for more proportional and, arguably, more representative participation.

Finally, governments are re-evaluating the mechanisms they use to govern. There is a growing interest in developing more creative solutions to regulatory problems and making sure the mechanism used is the best suited to meeting a particular objective. Governments, realising the limits of traditional command and control regulation, seem more willing to explore less traditional regulatory techniques such as self-regulation, privatisation, economic instruments and contractual arrangements.

There are many reasons for studying these changes in the context of policies aimed at either exploiting or protecting the environment. Policy makers in these areas are heavily dependent on science and technology for determining the best solution to a particular problem and this science is continually being developed. Environmental problems tend to be long term problems in need of long term solutions. Furthermore, environmental issues tend to be global whereby the acts of one region may impact on another. As the environment affects everyone, environmental policy is becoming a concern of more and more people. The agreed global solutions often need to be accepted and implemented locally in order to be successful. Also, with increased commitments to sustainable development, environmental concerns cannot simply be 'bolted on' at a late stage in decision making and instead must be treated alongside and weighed against their economic and social counterparts.

This issue of *Hume Papers on Public Policy* is devoted to examination of environmental policy in the UK, especially in Scotland. In particular, it examines the changing objectives used to justify public action on behalf of the environment, some of the mechanisms used to meet these changing objectives and certain institutional changes both internal and external to these developments.

The first paper by Ross considers the theory behind why states regulate on behalf of the environment. She notes that state intervention is usually justified as necessary to meet some public interest goal that the market if left unchecked would overlook. The paper begins with an examination of why the market together with the private law cannot adequately meet public expectations about environmental protection. It then compares the traditional justifications given for environmental regulation namely, correcting market deficiencies such as inadequate information, externalities and redistributing wealth with more recent calls for intervention based on concepts such as environmental justice and sustainable development. Ross concludes that while most regulation designed to fulfil some element of environmental justice can also be justified by one or more traditional market based goals, the converse is not true. The market based goals are capable of reflecting any number of different ideologies from socialist to conservative. In contrast, certain ideological values that favour environmental protection are inherent in environmental justice. Thus, there is no need for policy makers to fear justifying environmental protection measures on environmental justice grounds since these are consistent with the traditional justifications aimed at correcting market deficiencies or redistributing wealth. The ideological shift in favour of sustainable development and environmental protection can be accommodated in the same way as socialist, republican or liberal values have been in the past. Indeed, it is the required ideological and cultural changes that are the real hurdles for those promoting environmental justice.

Once collective action has been justified the question then becomes what form of action is most appropriate. The next two papers in this volume consider the relationship between the objectives of certain environmental regimes and the mechanisms they employ.

Last's paper on the evolution of habitat protection in the UK begins with an historical account of the reasoning behind the initial intervention by the state for the purpose of habitat protection. She highlights how the objectives for habitat protection have developed from simply protecting sites from the dangers of large-scale development to a more integrated approach to eco-system management. She argues that as the objectives have evolved, the mechanisms used to meet them have also had to change. For example, the increased reliance on co-operation between countryside agencies and landowners reflects moves towards a more integrated and sustainable approach to the countryside. Last observes that the apparently ecocentric objectives in the proposed legislation for England and Wales differ dramatically from those in the Scottish proposals which continue to be anthropocentric. She then argues that the mechanisms proposed in each jurisdiction reflect this divergence and that devolution in Scotland may further exacerbate matters.

Of the new styles of regulation being put forward to meet changing objectives some of the most popular fall under the heading of economic instruments. In their paper Hanley and MacMillan look at the use of economic instruments such as financial incentives, environmental taxes and tradeable entitlements as a means of influencing land use and contrast them with more conventional 'command and control' style regulation. Hanley and MacMillan examine four specific examples in Scotland where economic instruments are used or could be used effectively: the management of red deer/salmon, congestion in national parks, reducing the impact of farming on wildlife and landscape, and encouraging multi-use forestry. They conclude that the best solutions to many problems often involve using economic instruments in combination with other approaches to regulation and that at present, the full potential of economic instruments is not being realised.

Changing the regulatory style may not be enough to meet certain policy objectives and some sort of structural or institutional change may be required. Throughout their paper on the development of the provision of water in Scotland, Sawkins and Dickie highlight the need for appropriate institutional structures to ensure not only the effective delivery of the service but also the effective protection of the resource. The paper begins with an analysis of the major legislative and institutional changes in the Scottish water industry since the Second World War, highlighting the conflicting objectives of providing a quality service, at a reasonable price while preserving environmental quality. In particular, they examine the role of the new Water Industry Commissioner for Scotland and his relationship with the water authorities, the Scottish Environment Protection Agency, and the Scottish Executive. They conclude that past problems such as short termism, a lack of accountability and the division of economic responsibilities appear to have been largely resolved. They see the main challenges for the Commissioner to be asserting his operational independence from the Scottish Executive and helping the water authorities find the right balance between keeping prices down and financing essential capital expenditure programmes.

Reid, in his paper, considers how devolution in Scotland will affect environmental protection. He begins by examining the legislative and administrative arrangements for environmental protection following devolution and then considers the progress made by the Scottish Parliament and the Scottish Executive so far. Reid observes that environmental protection across the UK has never been truly uniform due to, among other things, the different legal systems. He further notes that the devolution settlement imposes considerable constraints on both the Scottish Parliament and the Executive's freedom to act. These constraints, including the requirement to observe EC law and the long list of matters reserved to Westminster, will limit any divergence between the laws in England and Wales and those in Scotland. That said, Reid concludes that devolution is likely to cause some divergence in the law on environmental protection in the UK citing the proposals on nature conservation as one example. He also concludes that the process appears to have slowed Scotland's progress both in terms of legislating and implementing laws on the environment.

The environment has recently gained fresh prominence at a time when governments are reassessing their roles and how best to achieve their policy objectives. Environmental regulation is at the forefront of broader shifts in public policy and decision making more generally. This volume of the Hume Papers on Public Policy is intended to add to the discussion in this area. It is important to emphasise that while the changes discussed are in the context of environmental protection, they are occurring in many other areas of public policy making.

Acknowledgements

The editor would like to thank all the contributors whose papers are published in this volume as well as Professor Brian Main at The David Hume Institute for arranging its publication.

Andrea Ross

Justifying Environmental Regulation

*Andrea Ross**

Introduction

Typically, state intervention in the market is justified on public interest objectives which tend to be confined to circumstances where intervention is needed to correct some deficiency in the market or to re-distribute a benefit or burden. Recently, there also has been pressure to intervene in the market in the name of environmental justice and other environmental principles. The aim of this paper is to consider the correlation between these two sets of public interest objectives. It begins by examining how the market with the support of both the criminal and private law allocates goods and services and how in certain circumstances this allocation is not in line with the public interest. Next, it specifically considers the drawbacks of the market system in securing environmental protection. The conventional 'public interest' grounds used to correct these deficiencies are then considered in the context of environmental justice.[1]

The Market System and Environmental Protection

The key mechanism used to allocate goods and services in the world today is the market. Given that different skills and resources are available to different people, individuals need to co-operate if they are going to satisfy fundamental human needs. Individuals become specialists in the various areas of production and will use the market to sell surplus goods and services and buy those they need. For example, it is easier for John to bake three loaves of bread instead of John, Ann and Bill all baking one loaf each. John can then sell the

* The author would like to thank Rowan Middleton for her superb research assistance and all those who have commented on previous drafts of this paper, in particular Elizabeth Kirk.

1 As this paper covers a wide range of material, it would be impossible to give a full account of every topic and issue it raises. Where possible, the author has directed readers to more detailed discussions elsewhere.

surplus loaves on the market in order to have money to buy other goods and services he needs. Thus, the market can be described as a series of utility maximising contracts between relevant individuals. The premise behind the market system is that individuals and groups are left free, subject only to very basic restrictions, to pursue their own welfare goals.[2]

The market model contends that almost all conceivable welfare goals can be pursued by individuals trading with each other. It is dependent on several key assumptions. First, individuals are assumed to act rationally by choosing courses of action which maximise their utility. Second, individuals are assumed to have all the information they need to make rational, utility maximising decisions. Third, the model assumes that social welfare is the aggregate of individual welfare and that 'the public interest' will be achieved by individuals acting in a self serving fashion. Furthermore, the model presupposes perfect competition and assumes minimal transaction costs.[3]

Over time it has become apparent that a non-interventionist approach to environmental protection which relies solely on market forces to direct individuals towards environmentally-friendly behaviour, is insufficient to produce the level of environmental protection the public expects. Bunyard notes "The pattern of consumption and behaviour have rested in the past on a notional right to exploit natural resources to further human progress and have reflected the urge to go on taking, to forgo restraint."[4]

The main reason a non-interventionist approach fails to protect the environment adequately is that the market relies on competition to allocate resources and yet there is no 'market' in a traditional sense for environmental amenities like drinking water, diverse eco-systems and clean air. Thus, they are exposed to over use by opportunists. Historically, environmental resources such as these have been treated as 'free goods' of which there is virtually an endless supply. A zero price leads to an increased demand for these 'free goods'. At the same time because they are free there is no incentive to protect them. This results in three distinct market failings: loss of efficiency, failure to account for collective action problems and distributional problems. First, the cost of using these free goods is not taken into account in market transactions and so the result is not efficient. For example, the cost of driving a car to work every day has a hidden environmental cost to the atmosphere. Thus, the decision to drive to work instead of bicycling to work may be inefficient because it fails to consider all of the costs of doing so.[5] Second, individually rational private behaviour may produce collective or public irrationality. For example, the cost of an individual car's emissions on the environment may be negligible however, the collective effect on the atmosphere of the national or world population of cars is substantial. In fact, we now know that the actual supply of environmental amenities

2 Ogus *Regulation Legal Form and Economic Theory* (Oxford, Clarendon Press, 1994) pp. 15–17; Coase "The Problem of Social Cost" (1960) Vol. III *J. Law & Econ.* pp. 1–44
3 Ogus *op. cit.* pp. 23–24.
4 Bunyard "Antartica: The ethics of Conservation" in Mayers (Ed), *Antartica: The Scientists' Case for a World Park*, (Greenpeace, 1991).
5 Pearce, *Blueprint 3 – Measuring sustainable development* (London, Earthscan Publications, 1993) p.5

like clean air is being diminished daily by polluting activities and that at some point in the future, the supply will no longer be capable of meeting demand. Once supply is affected this can lead to the third market failing, distributional problems. A non-interventionist would argue that a diminished supply will in time lead to the creation of a market for these amenities. This is already true for some fishing grounds and for drinking water in some countries where it must be bought in bottles. The market approach to allocation may result in the clean air or water being distributed in what many people would consider to be an unacceptable manner (to the highest bidder). Taken to its logical extreme, this would mean that wealthier states, individuals and species would end up with access to the necessities of life at the expense of the rest.[6]

The Role of the Private Law in the Market System

The legal system supports the operation of the market.[7] Individual rights are protected by the criminal law and by the private laws of property, tort and contract.[8] For example, the criminal law protects certain inalienable rights such as personal integrity through the laws relating to murder, theft and assault. It also penalises those who steal or vandalise another's property. The private law protects owners from the unlawful interference with the use and enjoyment of their property and the private laws of contract and tort are used to facilitate market transactions by giving effect to bargains and ensure fair dealing. The private law has also developed to deal with some of the failings in the market system and to ensure it operates effectively and efficiently. For instance, if a market transaction harms the private property rights of a uninvolved third party then that party has a right to be compensated for their loss for example, by way of an action in nuisance.

The private law has always served as a vehicle for environmental protection.[9] However, it does so primarily through the control of competing interests in land rather than any goal of protecting the environment for its own sake.

---

[6] Pearce, Markandya, Barbier, *Blueprint for a Green Economy* (London, Earthscan Publications, 1989) pp. 4–5. *ibid.* p.5

[7] This section largely refers to the role of English Common Law in relation to the market system. However all market economies are reliant on support from the relevant legal system. Note the allocation of rights in a given legal system will profoundly affect the distribution in the market. See Calabresi, and Melamed "Property Rules, Liability Rules and Inalienability: One View of the Cathedral" (1972) *Harvard Law Review* Vol. 85 No.6 pp 1089–1128 and below in Chapter 3.

[8] Coase *op.cit.* p. 8. "It is necessary to know whether the damaging business is liable or not for damage caused since without the establishment of this initial delimitation of rights there can be no market transactions to transfer and recombine them."

[9] See generally, Bell, *Ball & Bell on Environmental Law* 4th Edition (London, Blackstone Press, 1997) Chaps. 3 and 8; Hughes, *Environmental Law* 3rd Edition (London, Butterworths, 1996) Chap. 2; Cameron "Civil Liability for Environmental Harm" Chap 9, in Reid (Ed) *Environmental Law in Scotland* 2nd Edition (Edinburgh, W Greens / Sweet and Maxwell, 1997).

Consequently, in some cases, where the individual's interests and that of the environment coincide, the private law can be a very useful tool in the armoury of environmental protection. For example, those in need of clean water such as whisky distillers in Scotland have historically relied on actions in nuisance to protect the quality and purity of water sources.[10] Indeed, as Bell points out there may be many instances where the remedies available through the common law will give a far wider range of options to potential plaintiffs than are available through seeking statutory bodies to act on their behalf.[11] For example, one of the reasons planning agreements are particularly attractive to planning authorities is because they can be enforced using contractual remedies as opposed to the more drawn out statutory procedures.[12]

Often however, the private law has proven to be an inadequate means of protecting the environment. The main reason is that, like the market, the private law is dependent on rights and is designed to protect those rights. The private law requires some harm to have come to a recognised interest. This has several consequences. First private actions are aimed at compensating the individual for loss suffered and as a result do not necessarily solve the environmental problem. Second, as the private law seeks to balance competing individual rights, the right to take action is normally only vested in those who are directly harmed: generally the individual in possession of land which suffers damage.[13] The transaction costs of bringing individual actions may be prohibitive and the scope for bringing class/group actions is limited.[14] Moreover, rational individuals and firms will only seek to enforce rights where the expected benefits (compensation, an injunction or interdict restraining the harmful activity) exceed the expected costs including time and legal fees.[15] Third, it does not adequately seek to compensate for environmental damage where there are no rights of ownership to protect. For many environmental harms, such as damage

---

[10] See for example *Young v Bankier Distillery Company* [1893] AC 691, *Ballard v. Tomlinson* (1885) 29 Ch.D 115, *R. v. Bradford Navigation Co.* (1865) 6 B. & S. 631 and more recently, *Cambridge Water Co. v. Eastern Counties Leather plc* [1994] 2 AC 264. The private law also provided the basis of the decision in the *Trail Smelter arbitration*, 99 *AJIL* (1939), 182; 35 *AJIL* (1941) 684 which is a landmark decision in modern international environmental law.

[11] Bell *op. cit.* p.203.

[12] Town and Country Planning (Scotland)Act 1997 s75. *Avon County Council v. Millard* [1986]JPL 557. See Purdue, Young, Rowan-Robinson *Planning Law and Procedure* (London, Butterworths, 1989) pp.285–287.

[13] Reid *Nature Conservation Law* (Edinburgh, W Green / Sweet & Maxwell, 1994) pp. 14–16.

[14] Class actions generally require all the members of the class to have suffered the same harm from the same cause. To date they have been used in only a very limited number of circumstances and are not available in many jurisdictions. See for example *Civil Procedure Rules 1999*. Generally see Robinson, Dunkley (eds.) *Public Interest Perspectives in Environmental Law* (London, Wiley Chancery, 1995) Wade, Forsyth *Administrative Law* 7th ed. (Oxford, 1998) pp. 601–607.

[15] Coase *op. cit.* p.17 "In the standard case of a smoke nuisance which may affect a vast number of people engaged in a wide variety of activities the administrative costs might

to a coral reef due to oil exploration, there may be no identifiable person with a direct interest.

The utility of the private law as a mechanism for dealing with environmental problems is also dependent on the remedies available once a cause of action has been established. In some instances these provide very real solutions.[16] However, often for the reasons cited above the remedies fall short of actually resolving the harm to the environment. There are three main types of remedies which are can be sought: damages, injunctions (interdict in Scotland) and specific performance (specific implement in Scotland).

The most common remedy following an incident involving harm to property such as a pollution spill would be an action for damages.[17] The aim of an award of damages is to place the plaintiff as far as possible in the position they would have been had the wrongful act not occurred. This can be calculated in two ways: on the cost of the clean up operations necessary to restore the property to its previous state or on the difference between the value of the property as it was after the pollution affected it and before. Damages for future loss are only available where an injunction is not granted and tend to be used sparingly. In England, exemplary damages can be awarded in specific instances to deter the defendant and others from committing torts which may result in financial benefit to the person responsible.[18] Furthermore, there is no obligation to use any awarded damages to rectify the harm.

The injunction or interdict in Scotland is a discretionary remedy which can be used to prohibit a defendant / defender from carrying on an activity which is causing harm. The activity complained of has to be continuing at the date of the action, or there has to be a threat that the activity will continue. In exercising its discretion, the court will consider whether or not the activity is of sufficient gravity or duration to justify stopping the defender's activities. The harm suffered to the pursuer has to be balanced against the effect that granting the interdict would have on the defender. In this regard, an interdict may be suspended or qualified.[19] An interdict may also be sought prior to the occurrence

---

[15] (continued) well be so high as to make any attempt to deal with the problem within the confines of a single firm [or individual (Ed)] impossible . . . Instead of instituting a legal system of rights which can be modified by transactions on the market, the Government may impose regulations which state what people must or must not do and which have to be obeyed."

[16] See planning agreements example and note above.

[17] See *Marquis of Granby v. Bakewell UDC* (1923) 87 JP 105, *Scott-Whitehead v. National Coal Board* (1987) 53 P &CR 263, *Cambridge Water Co v. Eastern Counties Leather plc* [1994] 2AC 264.

[18] These are limited to certain classes of tort and cases where there has been oppressive, arbitrary or unconstitutional action by servants of the Government, or where the defendant stands to make a profit from their conduct which surpasses any award of damages. See *A.B. v. South West Water Services Ltd* [1993] QB 507.

[19] In *Webster v. Lord Advocate* 1985 SC 173 the interdict prohibiting the setting up and dismantling of scaffolding for seating for the Edinburgh tattoo was suspended for 6 months to allow the organisers time to find a quieter alternative. See also *Halsey v. Esso Petroleum Co Ltd* [1961] 2All ER 145.

of the event causing damage or injury. However, there must be sufficient proof of imminent damage and it must be demonstrated that if the activity were to continue the damage accruing would be substantial and of such degree that it would be difficult to rectify. This type of injunction can be used to prevent harm to the environment. For example, injunctions have been used to prevent continued pollution to watercourses[20] and could be sought to prevent proposed activities which would have irreversible environmental effects such as damming or diverting rivers.

An order for specific performance (specific implement in Scotland) is one by which the courts direct the defendant to perform a contract in accordance with its terms. At first glance, this remedy appears to have a great deal of potential in the context of environmental law. It could be used to force a party to adhere to the terms of a supply contract imposing obligations on the environmental quality of goods supplied (timber from renewable sources) or in regard to a planning or management agreement imposing monitoring or restoration obligations.[21] Specific performance is however, a discretionary equitable remedy which is not available as a matter of course. As a general rule, an order for specific performance will not be made against a defendant in any case where damages are an adequate and appropriate remedy, where, had the positions been reversed, the plaintiff's undertaking could not have been specifically enforced, or where the contract has been discharged or is no longer in existence. Beatson suggests that while the courts appear to be taking a more liberal approach to the availability of specific performance, it should remain a secondary remedy to damages.[22] Moreover, not all contractual undertakings, particularly those to perform services or to build are sufficiently precise to be enforced specifically.[23]

The criminal law has its own limitations. Environmental crimes are often not perceived as 'real' crime and resort to the criminal law is often seen as heavy handed especially when the harm has been caused inadvertently or accidentally.[24] As a result, few environmental crimes are prosecuted and when prosecuted the full penalty available is rarely imposed. Consequently, it is then easy for companies to see fines as part of the cost of doing business as they do not unduly interrupt business nor impose a real burden on it.[25] The

---

[20] *Young v Bankier Distillery Company* [1893] AC 691.

[21] See for example under the Town and Country Planning (Scotland) Act 1997 s. 75 and under the Countryside Act 1968 s. 15.

[22] Beatson *Anson's Law of Contract* 27th Edition (Oxford, Oxford University Press, 1998) at p. 596. This is because specific performance avoids the policy of the mitigation rule (plaintiff should keep losses to a minimum) and many more losses can now be identified and quantified in monetary terms.

[23] Beatson *op. cit.* p. 414.

[24] Rowan-Robinson, Watchman, Barker, *Crime and Regulation*, (Edinburgh, T &T Clark, 1990) p.297.

[25] For low fines see "Low fines fail to deter waste offenders" ENDS Report 226 November 1993. In contrast, Milford Haven Port Authority was fined £4 million for its role in the Sea Empress oil pollution disaster off Pembrokeshire in February 1996. ENDS Report 288 p.50.

issue surrounds the aim of using the criminal law for environmental protection. Is it to impose the true cost of polluting on the polluter or is it to prevent the damage in the first place?[26] If an element of public embarrassment, loss of reputation or custom accompanies the criminal sanction then the penalty may be a useful deterrent. Furthermore, if part of the penalty includes paying for the cost of cleaning up the pollution then not only does the penalty achieve the above, it also actually helps solve the environmental problem. For example, the fine imposed on Shell of £1 million for polluting the River Mersey has received wide publicity. It acted as a meaningful deterrent not only in monetary terms but also due to the publicity. Less publicity has been given to the fact that Shell are reported to have paid £1.4 million in clean-up costs.[27] A requirement to pay for loss arising from a breach of control could be of far greater consequence to an industrial operation in terms of financial penalty than any fine imposed by the magistrates and it helps repair the damage done.

The traditional system of adjudicating and enforcing rights introduces further inherent restrictions for both the criminal law and private law. The courts normally only have jurisdiction to enforce the law after harm has occurred and damage has been inflicted. The system is not well suited at preventing harm from occurring. Instead, it tends to react to events that have already occurred. Actions for injunctions or interdicts and instances where the terms in a contract may alter one party's behaviour so that harm is prevented are two possible exceptions.[28] Also, while the private law compensates the victim of the harm and the criminal law penalises the wrongdoer neither system actually resolves the environmental problem or cleans up the damage caused.[29] Furthermore, as the courts rely on imprecise standards of conduct under the heading of 'reasonable', the law is regressive. Often the areas most in need of protection, fail to receive it because the activity in question may be reasonable in that particular area. Thus, the most polluted areas may receive the least assistance.[30] Finally, it is often very difficult in an environmental case to obtain the evidence necessary to establish a causal link between the origins of the harm and the actual damage caused. There may be ten factories within twenty miles of each other, all dispersing similar acidic emissions into the atmosphere. To succeed the person bringing the action must be able to show who caused how much damage, yet

---

[26] See Rowan-Robinson, Ross "Enforcement of Environmental Regulation in Britain: Strengthening the Link" (1994) *JPEL* pp.200–218.
[27] *National Rivers Authority v. Shell UK Ltd* [1990] Water Law 40.
[28] These have their own difficulties. Injunctions are difficult to obtain and seen as heavy handed in some instances. (see above) Preventative contract terms are limited to those situations where a contract exists and often are limited to protecting the interests of signatories. See Ross, Rowan-Robinson "Environmental Information and the Greening of Industry" (1997) *JEPM* 40 (1) pp.111–124 at pp.116–118.
[29] Although the amount of damages awarded may be linked to the cost of cleaning up the land, there is no guarantee that a successful litigant will use the money for this purpose.
[30] *Sturges v. Bridgman* (1879) 11 ChD 852: "What would be a nuisance in Belgrave Square would not necessarily be so in Bermondsey".

this is a very technical question requiring detailed scientific investigations. In practice, this information may be very difficult and extremely costly to obtain. These problems are compounded in a criminal trial where the standard of proof is higher. These difficulties may again discourage individuals and in the case of criminal law, the state, from taking action.[31]

Environmental Justice

The want of rights and frailty of protection under the law described above, have resulted in increased pressure at all levels of policy making to give the environment some form of political recognition.[32] One solution has been a call for increased environmental rights either through the human rights regimes or independently. Principle 1 of the 1972 Stockholm Declaration[33] declares that "man has the fundamental right to freedom, equality and adequate conditions of life in a environment of a quality that permits a life of dignity and well being" and seems to support the idea that an individual right to a decent environment exists. Boyle[34] notes that this initial emphasis on a human rights perspective has not been maintained and that the Rio Declaration avoids the terminology of rights altogether and merely asserts that: "Human beings are at the centre of concerns for sustainable development. They are entitled to a healthy and productive life in harmony with nature."[35] A 1990 UNESCO study argued that a collective right to a decent environment, similar to the right to self determination, exists and forms an essential basis for the full enjoyment of individual rights, however, this also remains controversial.[36] Another approach is to derive environmental rights from other human rights such as life, health and property.[37]

---

[31] Rowan-Robinson, Watchman, Barker, *Crime and Regulation*, (Edinburgh, T &T Clark, 1990); Richardson, Ogus, Burrows *Policing Pollution* (Oxford, Clarendon Press, 1982); Hawkins *Environment and Enforcement* (Oxford, Clarendon Press, 1984).

[32] Generally see proceedings from W.G. Hart Workshop on Access to Environmental Justice July 1997 London and Gillespie *International Environmental Law Policy and Ethics* (Oxford, Clarendon Press, 1997).

[33] Declaration on the Human Environment, Principle 1, *Report of the United Nations Conference on the Human Environment* (New York, 1973), UN Doc. A/CONF.48/14/ Rev.1, adopted in UNGA Res 2997 (XXVII) of 1972.

[34] Boyle, "The Role of International Human Rights Law in the Protection of the Environment" in Boyle, Anderson (eds.) *Human Rights Approaches to Environmental Protection* (Oxford, Clarendon Press, 1996) p.43; Birnie, Boyle *International Law and the Environment* (Oxford, Clarendon Press, 1992) p.191.

[35] Declaration on Environment and Development, Principle 1, *Report of the UN Conference on Environment and Development*, (New York, 1992) UN Doc. A/CONF.151/ 26/Rev.1.

[36] UNESCO, Final Report and Recommendations of an International Meeting of Experts, 1989 11 *HRLJ* (1990) 441. Boyle *op.cit.*

[37] See for example Article 12 of the 1966 UN Covenant on Economic and Social Rights which refers to the right to improvement of environmental and industrial hygiene.

All of these human rights approaches, however, only attribute worth to the environment and its natural resources to the extent that they benefit humans. Birnie and Boyle observe that some international and national laws do recognise the intrinsic value of the environment[38] and conclude that these tend to suggest the formulation of a collective right to a decent environment which allows many interests to be balanced through international co-operation and supervisory institutions. This may be difficult to achieve in practice as these institutions must take a holistic view capable of looking beyond human rights.[39]

Finally, the right to a decent environment need not be seen as a substantive collective right, but instead could be used to secure individuals rights of access to information, to participation in decision making processes and to administrative and judicial remedies. This procedural interpretation means rights can be exercised on behalf of the environment or its non-human components. This approach to the right is reflected both nationally, and internationally through, for instance, the extension of the laws of standing to include interest groups,[40] and legislation improving public access to environmental information.[41]

Restricting environmental rights to procedural rights while easy to accept still fails to attribute substantive value to the environment and its resources. One solution, and the one favoured by the author, has been to avoid the language of rights and instead speak more broadly in terms of environmental justice. Environmental justice tends to be defined as a broad political aim encompassing rights, goals, policies and principles.[42] The pursuit of justice is a strong public interest goal. Bryant defines environmental justice as "Broader than environmental equity. It refers to those cultural norms and values, rules, regulations, behaviors, policies and decisions to support sustainable communities, where people can interact with confidence that their environment is safe,

---

[38] See Biological Diversity Convention 31 *ILM* (1992), 822; the 1991 Protocol on Environmental Protection to the 1959 Antarctic Treaty 30 *ILM* (1991) 1455. Also national legislation such as in the UK the Environmental Protection Act 1990. See generally, Redgwell "Life, the Universe and Everything: a Critique of Anthropocentric Rights" Chap 4. of Boyle, Anderson (eds.) *Human Rights Approaches to Environmental Protection* (Oxford, Clarendon Press, 1996).

[39] Birnie, Boyle *op. cit.* p.194.

[40] See *Sierra Club v. Morton*, 405 US 727 (1972), *Australian Conservation Foundation v. Commonwealth of Australia* (1980) 54 ALJR, *Environmental Defence Society v. South Pacific Aluminium (No.3)* (1981) 1 NZLR 216 *R v. Her Majesty's Inspectorate of Pollution ex parte Greenpeace Ltd. (No.2)* [1994] 4All ER 329, *R v. Secretary of State for Foreign Affairs ex parte World Development Movement Ltd* [1995] 1 All ER 611.

[41] See for example EC Directive 90/313 on Access to Information on the Environment and in the UK the Environmental Protection Act ss. 64–67; UNECE Convention on Access to Information, Public Participation in Decisionmaking and Access to Justice in Environmental Matters, Aarhus, June 1998 ECE/CEP/43/Add.1/Rev.1.

[42] In some of the US literature, environmental justice focuses narrowly on environmental racism. See for example Bullard (ed) *Unequal Protection: Environmental Justice and Communities of Color* (Sierra Club, 1994).

nurturing and productive. Environmental justice is served when people can realize their potential without experiencing the 'isms'.[43] Environmental justice is supported by decent paying and safe jobs; quality schools and recreation; personal empowerment; and communities free of violence drugs and poverty. These are communities where both cultural and biological diversity are respected and highly revered and where distributive justice prevails"[44]

Bryant's definition of environmental justice sounds like utopia however, it is designed to be that way. Grant takes a similar perspective. He maintains that environmental justice includes but goes beyond conventional legal justice, which protects individuals from the state through the use of institutions and processes such as the right to be heard, the right to counsel and the right to be given reasons. Environmental justice includes broader questions of political justice and morality. It includes distributional justice and takes an ethical perspective. Grant submits that this is evidenced by the emergence of principles such as sustainable development, the polluter pays principle and the precautionary principle which are to underpin future environmental law and policy.[45]

Indeed the environmental principles have emerged as components of environmental justice. The most important of these include: the polluter pays principle, the precautionary principle, the principle of shared responsibility and sustainable development. These are described below.

The Environmental Principles underpinning Environmental Justice

Polluter Pays Principle
Perhaps the most straight-forward of the environmental principles is the notion of making the polluter pay. It is widely accepted that the costs of environmental damage caused by polluting activities should be borne by the person responsible for such pollution including the cost of administering any regulatory scheme and paying for any clean up operations.[46] Making the party at fault for any damage as opposed to innocent victims or society as a whole, pay for the costs of that damage is a valid public interest goal. However, while widely accepted at all levels of policy making, it is often difficult to actually make the polluter pay as often the polluter can not be identified due to the passage of time or the number of potential polluters. In the case of historic pollution such as contaminated land the polluter may no longer exist. Moreover, the costs attributable to a given polluter may be difficult to ascertain. For example, while it is accepted that air pollution gives rise to increased incidences of asthma, it would be hard to identify all the health and economic costs attributable to a given polluter.

---

[43] Racism, sexism, ageism etc.

[44] Bryant *Environmental Justice: Issues, Policies and Solutions*, (Washington, Island Press,1995) p. 6

[45] Grant (1997) Faculty of Advocates November 1997.

[46] European Union Fifth Action Programme on the Environment (1992) *Towards Sustainability* COM (92)23; HMSO *This Common Inheritance* 1990 Cm 1200.

Precautionary Principle
The precautionary principle provides that effort should be made to prevent harm from occurring instead of only reacting to disasters after they occur and that precautionary action should be taken where there are significant risks to the environment even when scientific evidence is not conclusive.[47] The problems surrounding the precautionary principle are twofold. First, in what instances should a precautionary principle be even considered? There is some debate as to whether it is appropriate for some activities. For example, development control in the UK is historically based on a presumption in favour of development which largely runs contrary to the precautionary principle.[48] Even when the decision maker is willing to consider a precautionary approach, there are issues as to how much scientific evidence is necessary to invoke it.[49] The continuous debates on whether there is sufficient evidence to take action to curb global warming and the subsequent argument on how much action is required illustrate the difficulties in applying the precautionary principle.[50]

Shared responsibility
Another key environmental principle is that of shared responsibility. The European Community's Fifth Action Programme on the Environment entitled *Towards Sustainability* advocated that responsibility for environmental problems should be shared between government, producers and consumers.[51] The realisation is that environmental improvement cannot simply come from the top down. Individuals and businesses must also be prepared to take action to improve the state of the environment in terms of their consumer choices, their political activities and in their daily lives.

Sustainable Development
The principle of sustainable development is the broadest of the environmental principles and to a large extent encapsulates the others. At the heart of the principle of sustainable development is the recognition that the world's environment, its economies and the way it treats its human and other

---

[47] European Union Fifth Action Programme on the Environment (1992) *Towards Sustainability* COM (92)23; HMSO *This Common Inheritance* 1990 Cm 1200.

[48] Walton, Rowan-Robinson, Ross 'The Precautionary Principle and the U.K. Planning System' (1995) *Journal of Environmental Law and Management* Vol. 7 Issue 1 pp.35–40.

[49] See for example Department of the Environment (1995) *A Guide to Risk Assessment and Risk Management for Environmental Protection* HMSO, London, which states at p.44 "The precautionary principle is not a licence to invent hypothetical consequences".

[50] For example contrast the views of Easterbrook *A Moment on the Earth* (New York, Penguin Books, 1995) Chaps. 16, 17 with Krause, Bach and Kooney "A Target-Based, Least Cost Approach to Climate Stabilization" Chap. 2 in Kirkby, O'Keefe, Timberlake (eds.) *Sustainable Development* (London, Earthscan Publications, 1995).

[51] COM (92)23; see also HMSO *This Common Inheritance* 1990 Cm 1200 at pp. 10, 16.

inhabitants both present and future are all inter-linked.[52] The most widely cited definition of sustainable development is provided in the Brundtland Report: "development that meets the needs of the present generation without compromising the ability of future generations to meet their own needs".[53] While this definition is attractive as it offers a comprehensive and consensual approach which is able to 'weld together quite disparate and conflicting interests', it suffers from being vague and imprecise. It can mean all things to all people.

Indeed, the differences between what Pearce et al. refer to as weak sustainable development and strong sustainable development are substantial.[54] Weak sustainable development is indifferent to how we pass on capital stock and environmental capital can be substituted for man-made stock. For example, a tropical forest may in certain circumstances be legitimately substituted by a new community with schools, homes and a medical centre. In contrast, strong sustainable development allows some substitution but defines certain environmental assets (clean air, perhaps a tropical forest) as critical to our well-being. Sustainability refers to the maintenance of environmental capacities of those critical assets over time.[55]

To date many developed nations in their discussions about sustainable development have focused on balancing environmental assets with maintaining economic prosperity. More recently, the social component of sustainable development has been coming to the forefront. This has always been a concern of the developing world but is only starting to appear in the national strategies of developed nations.[56] In its white paper, *A Better Quality of Life – A strategy for sustainable development for the UK*[57] the Government includes social progress which recognises the needs of everyone as one of its four objectives for its vision of sustainable development along with: the maintenance of high and stable levels of economic growth and employment; effective protection of the environment and the prudent use of resources. Achieving sustainable development means addressing all of these objectives equally, for both present and future generations.[58]

Environmental Justice and the Environmental Principles as Public Interest Goals

Sustainable development and environmental justice have been intentionally framed in vague terms in order to increase acceptance among divergent

[52] 70 different definitions are listed in Holmberg, Sundbrook "Sustainable Development: What is to be Done" in Holmberg (Ed) *Policies for a Small Planet* (London, Earthscan, 1992) pp 19–38.
[53] World Commission on Environment and Development (the Brundtland Commission) *Our Common Future*, (Oxford, Oxford University Press, 1987) p. 43.
[54] Pearce, Markandya, Barbier, *Blueprint for a Green Economy* (London, Earthscan Publications, 1989) Chap.2.
[55] *ibid.*
[56] See Kirkby, O'Keefe, Timberlake, *The Earthscan Reader in Sustainable Development*, (London, Earthscan, 1995) Chap. 1.
[57] Cm 4345 (London, HMSO, 1999).
[58] Ibid para. 1.2.

political, economic and cultural groups.[59] Although there is widespread agreement that these words do indeed constitute principles of environmental law, there is much less agreement on, what precisely they mean. For instance, an anthropocentric approach to environmental law would accord the welfare of humanity as the primary rationale for any regime of environmental protection. In contrast, recent legislation[60] seems to reflect a shift towards a more ecocentric approach whereby natural eco-systems are seen as having an intrinsic worth, irrespective of the benefits they provide humans.[61] The difficulties with environmental rights discussed above resurface here. However, as the rights issue is only one component of environmental justice, it is possible to look beyond rights to see the bigger ideal. This approach is still problematic as many of the environmental principles are fraught with difficulties and the principles often conflict with one another. For example, making the polluter bear the brunt of any environmental costs may have undesirable social, economic or even environmental implications and hence, be unsustainable.

Increased public awareness about the state of the environment combined with the broad definitions attributed to environmental justice and its component principles have over time led to their acceptance at every level of state action. They were the focus at the Earth Summit held in Rio de Janeiro in 1992.[62] They are the foundations of European Community law on the environment as set out in the Fifth Action Programme on the Environment and are the basis of much national environmental policy.[63] More recently, the International Court of Justice has held that 'sustainable development' is now to be considered a 'norm' of international environmental law.[64] As such, these principles could legitimately be seen as reflections of the 'public interest'. In the context of this paper, it can be argued that they could be used to justify regulation or collective intervention in the market.

---

[59] For example, some people believe we should ensure a sustainable population of whales for the sake of communities who rely on whaling and whale products for their survival focusing on the management of the whole eco-system Others justify similar action because they question the humaneness of killing whales or to ensure future generations of humans can enjoy the beauty of whales. Still others wish to protect the whales for the sake of the whales themselves. See Gillespie *op. cit.* pp. 45–47

[60] See Environmental Protection Act 1990 sections s1–4.

[61] That said, the only way these can be protected from certain harms is for them to be valued by humans. We may value the ecosystem for its own sake but we (humans) still have to value it. For a fuller exploration of related issues see Gillespie *op.cit* Chap.9.

[62] Declaration on Environment and Development, *Report of the UN Conference on Environment and Development,* (New York, 1992) UN Doc. A/CONF.151/26/Rev.1.

[63] European Union Fifth Action Programme on the Environment (1992) *Towards Sustainability* COM (92)23; HMSO (1999) *A Better Quality of Life – A strategy for sustainable development for the UK* Cm 4345.

[64] Gabcikovo/Ngaymaras Dam decision, 1998 37 ILM 162 para 140.

Public Interest Theory and Environmental Justice

Ogus argues that where market failure is accompanied by 'private law' failure a prima facie case can be made on public interest grounds for collective intervention whereby the state intervenes to encourage or direct behaviour that (it is assumed) would not occur if the market was left to itself.[65] For example, collective intervention may be required where a market transaction harms an interest which is not recognised by the private law, such as the habitat of an endangered species. The term 'regulation' is used to describe the laws which implement this intervention. In contrast to the laws used to support the market system, these laws tend to be enforced by the state and as a result are often described as directive, public and centralised,[66] however, regulation can take many forms. Collective goals are often pursued by private agreements such as management agreements used in nature conservation, and through self-regulatory mechanisms which are far from centralised, public or directive.

The remainder of this paper is devoted to examining the 'public interest' goals commonly used by regulatory theorists to justify regulation in the context of environmental justice. Specifically, it considers how these goals relate to the environmental principles behind environmental justice. The focus is on those goals used to determine *whether* any intervention is warranted. Those public (and private) interest goals used to prefer one form of intervention over another are not examined here.

Public interest theories of regulation attribute to legislators and others responsible for the design of regulation a desire to pursue certain collective goals. The public interest goals justifying collectivist action can take many forms and will vary according to time, place and societal values.[67] The 'public interest theory' of regulation broadly defined[68] has two categories of 'public interest' goals which can be used to justify regulation: economic and non-economic public interest goals. The economic goals focus on achieving the most efficient allocation of resources by ensuring resources are put to their most

---

[65] Ogus *op. cit.* p.28 Ogus notes however, market failure and private law failure alone, are not sufficient grounds for taking regulatory action. There are instances where regulatory intervention may not be any better suited to resolving the problem. For example, the administrative costs of intervention may be prohibitive or the regulation may act as a disincentive or create a more severe inequity. As such, the consequences of any regulatory measure must also be thoroughly examined. This is dealt with in subsequent chapters.

[66] Ogus *op. cit.* p.3.

[67] Questions surrounding who represents the 'public interest' and how is it best determined are beyond the scope of this paper. However, it is important to recognise the difficulties surrounding the determination of what is in the public interest. Ogus notes that the study of motivation is an elusive and perhaps impossible task. Law is the result of debate and negotiation and there are often conflicting expressions of what is intended. Even when intention is clear how do we know these were not motivated by private interests.

[68] Narrow definitions tend to focus solely on the economic goals.

valued use. Generally, these aim to correct deficiencies in the basic assumptions upon which the market system is based. For instance, regulation may be required to create a market where there is no market to allocate a particular good (these are often called public goods) or where there is no information about a particular good. Non-economic public interest goals, in contrast, tend to focus on wider social, political and moral issues such as the furtherance of justice, a need to protect individuals from harm and altruistic motives.

Economic Goals

The market system, assuming complete and accurate information, full competition and no transaction costs, operates to generate allocative efficiency.[69] The allocation of resources is efficient when it is impossible to make any one individual better off without, at the same time, making someone else worse-off.[70] In real life however, information is often costly and / or deficient, competition is limited, transaction costs are high and the market solution in fact makes others worse off.[71] Sometimes the private law can act to resolve these types of market failure seeking compensation for those whose interests have been adversely affected. Often however, as discussed above the private law can not resolve the problem and in these instances, collectivist action may be required.[72] The main economic goals which may be used to justify regulatory intervention for environmental protection are: the regulation of externalities, the maintenance and protection of public goods (goods without a market), the improvement of information available, the resolution of co-ordination problems and solutions directed at exceptional market conditions.[73]

Externalities
Externalities arise when a producer's activity imposes costs or benefits on third parties that are not reflected (or internalised) in the price charged for the product resulting in a misallocation of resources. Purchasers of the product do not pay for its true social cost and as a result more or less units of the product are supplied than is socially appropriate.[74] Externalities can be positive or negative.

---

[69] Coase *op.cit.* The concept of allocative efficiency is problematic as a normative goal for economic welfare. See Ogus *op. cit.* p.24 and note directly below.

70 This is known as the *Pareto* test. It is criticised as being too narrow as one individual can veto a change which would benefit the rest of society. A broader alternative is the *Kaldor-Hicks* test which provides that a policy is efficient if it results in sufficient benefits for those who gain such that potentially they can compensate fully all the losers and still remain better off.

[71] Breyer "Typical Justifications for Regulation" Chap. 1 in Breyer *Regulation and its Reform* (London, Harvard University Press, 1982).

[72] See discussion of market system and private law above.

[73] It is very difficult to come up with an example of environmental regulation which is justified predominantly by a desire to avoid some monopoly. As this type of market failure is rarely associated with environmental regulation, it requires no further examination.

[74] Ogus *op. cit.* p.35.

For example, if one landowner increases the forest cover on her land, her neighbour (a gamekeeper) may benefit from an increased population of deer moving onto his land and an improved prospect of attracting shooting clients. In these instances, so long as the gamekeeper does not contribute to the afforestation, the positive externality or spillover benefit to the gamekeeper is not reflected in the overall cost of planting trees.[75]

Here, the focus is largely on negative externalities as pollution is the typical example of a negative externality. The owner of a waste disposal company may come to an agreement with a local farmer which permits the disposal of waste on some of the farmer's land. The agreement is unlikely to take into account the loss of amenity to neighbours and the local community. The waste heaps may be unsightly and smelly, and leaching waste could possibly contaminate neighbouring land. One solution would be to enter into agreements with all those third parties affected, however, the transaction costs of this often make it prohibitive. Furthermore, one third party may try to avoid paying and yet still receive the benefit (a freerider). Some externalities such as the one above could be resolved through the private law. Affected third parties could commence actions in nuisance seeking compensation for the harm they have suffered in the use and enjoyment of their land, however, the transaction costs of bringing individual actions may be prohibitive and the scope for bringing class/group actions is limited.[76] Moreover, rational individuals and firms will only seek to enforce rights where the expected benefits (compensation, an injunction restraining the harmful activity) exceed the expected costs including time, legal fees, information costs and trouble and so many legitimate claims are not pursued.[77] Collective action through regulation provides an alternative means of internalising the costs of externalities. In the waste disposal example, a regulation could be introduced which requires the company to obtain a licence for waste disposal. The licence may restrict the amount and type of waste which may be disposed of on the site and breach could result in a fine or a loss of license. Alternatively, the disposal company could be taxed on the amount of waste disposed of.

Both the private law solutions and the regulatory solutions can be seen as manifestations of the 'polluter pays principle'. However, while regulation to restrict emissions actually prevents the deterioration of the environment and maintains the business as a going concern, the private law can either compensate for the harm caused by the externality or if an injunction is sought shut down the business altogether.[78] The first private law solution often fails to achieve an environmentally just result since pollution may affect more than

[75] Breyer *op.cit.* Breyer notes that spillover benefits have sometimes been thought to justify government subsidy, as when free education is argued to have societal benefits far exceeding the amount which students would be willing to pay for its provision. Similar arguments might be made for subsidies to buy environmentally friendly equipment, technology or even for negative compensation schemes such as set aside.

[76] See note above.

[77] Ogus *op. cit.* p.27.

[78] Although, injunctions can be qualified so for example the waste disposal may be restricted from operating at certain times of day or night or on hot summer days.

those who have a right to compensation. Those affected yet uncompensated may include other communities, future generations and the particular environment itself. The second private law solution – the injunction / interdict – may be considered heavy-handed and inappropriate given the broader social and economic effects of closing a business. Thus, regulation to correct an externality may be more in line with a preventative and precautionary approach.

Pollution often involves irreversible ecological changes, which have no impact on those creating the harm and yet may adversely affect other countries, other species or future generations. For example, acid rain often affects lakes and forests miles away from the source of the pollution. With advancing scientific knowledge, increased environmental awareness and improved communications leading to a greater respect for our fellow inhabitants on the earth, the catalogue of environmental 'public bads' continues to expand. An example is the discovery that CFCs are destroying the ozone layer. This type of externality can cause problems for policy makers as it means trying to alter previously acceptable behaviour. Similar difficulties surround attempts to reduce our reliance on the private car to curb future climate change. Furthermore, environmental justice demands that we not only consider our closest neighbours but also those thousands of miles away and those living hundreds of years in the future. These affected groups often have no voice. The misallocation that results from an inability to consider such groups can not be resolved through private law mechanisms since the private rights will only accrue in the future. In such instances, regulation may be justified to prevent these externalities from harming present and future generations. Often there are uncertainties as to the extent of the harm (such as climate change) and the capacity of future technology to deal with the harm (nuclear waste). Any regulation aimed at correcting these types of externalities is consistent with a preventative principle, a precautionary principle and moves towards sustainable development. However, the difficulties surrounding how such regulation should proceed, can not and should not be minimised. For instance, how is the appropriate level of intervention to be determined given an infinite future population?

Sometimes an externality should not be corrected. For instance, while works to improve public transport links around a shop, may result in an initial loss of custom, following the completion of the works the shopkeeper may benefit from an increased amount of custom. In this situation most people would agree that the market adequately deals with the externality. There are other circumstances, however, where those concerned about the effect of the externality on the environment as a whole as opposed to its effect on one market actor would argue that the externality be corrected regardless of any benefits accruing. For example, if Sam buys a house next to a known contaminated site, the price he pays probably reflects the possibility of contaminants leaching onto his property.[79] Solely in terms of efficiency, it would be

---

[79] If the vendor is the builder who knowingly built on a suspect site then the same logic applies to the vendor. However, if the contamination was only discovered (due to increased awareness or new technology) after the vendor of the house had bought it, then

inappropriate to compensate Sam as he will have already received compensation for the harm through the reduction in the price paid. While this is the efficient solution, it does not deal with continued harm to the environment and in this instance is not environmentally just as it conflicts with the precautionary principle, the preventative principle and the concept of sustainable development.

Many externalities are ignored on the grounds that while an externality may give rise to a misallocation, the administrative and other costs of correcting it may outweigh the immediate social benefits arising from such action.[80] It is however, important to remember that, especially in the context of environmental protection, what may be a trivial cost to each individual affected by an activity, may in aggregate (and/or in the future) involve substantial costs. In these circumstances a substantial investment to regulate the conduct may be warranted. Thus, before choosing the regulatory solution it is important that the externalities of the regulation itself be measured and assessed. The key is to ensure all of the costs and benefits are 'integrated' into the decision making process. Moves towards imposing balancing obligations on decision-makers and subjecting new regulation and policy to appraisals, risk assessment and cost benefit analysis go some way in assessing the 'appropriateness' of a regulatory solution.[81]

Public Goods

Public goods are commodities the benefit from which is shared by the public as a whole, or by some group within it. No market exists to allocate this type of good. Consumption by one person of a public good does not leave less for others to consume and it is impossible or too costly for the supplier to exclude those who do not pay from the benefit.[82] As a result willingness to pay cannot be used to measure demand. Clean air and water have traditionally had the characteristics of public goods. As Sunstein explains the social costs of a polluting activity may dwarf the social benefits, but the costs are so diffused, and so small in the individual case, that the market will not force individual polluters to take those costs into account. Each polluter will continue, quite rationally, activities that will make society as a whole worse off. Due to the individual high costs and low benefits of seeking redress, victims will be slow to commence legal proceedings and instead will attempt to free ride on the remedial efforts of others.[83] The likely outcome is that neither the polluter nor any of its victims will limit the harmful effects of the activity, which will therefore be far higher than that which the environment can withstand.

---

[79] (*continued*) the vendor is bearing the cost of the contamination by having to accept a reduced price. This example demonstrates the difficulties in creating a simple solution to a problem like contaminated land. See Environment Act 1995 Part II.

[80] Coase *op. cit.* p.18.

[81] For a general discussion see papers by Last; and Hanley and MacMillan below.

[82] Ogus *op. cit.* p.33.

[83] Sunstein, *After the Rights Revolution: Reconceiving the Regulatory State* (Cambridge MA, Harvard University Press, 1990) at p. 49.

Consequently, to the extent that they are protected at all, most environmental public goods such as wild areas are protected by the state or by a state actor. By recognising that some environmental resources are public goods and acting to protect them from over exploitation the state is often acting in a preventative manner and taking precautionary action. Regulation may be the only way of ensuring that a particular resource is sustainably used and maintained.

International environmental law provides an interesting twist on the concept of public goods. Many international environmental concerns clearly fall within the notion of public goods.[84] The 1985 Vienna Convention for the Protection of the Ozone Layer[85] treats the whole stratospheric ozone layer as a global unity, without reference to the usual legal concepts of sovereignty, shared resources or common property. Rather it creates a new status for the global atmosphere which makes it appropriate to view the ozone layer as part of a common resource or common interest, regardless of who enjoys sovereignty over the airspace it occupies. Global climate change was similarly declared the 'common concern of mankind' in UN General Assembly Resolution 43/53.[86] Common concern is used to indicate the common legal interest of all states in protecting a resource whether they are directly injured or not and in enforcing rules concerning its protection.[87] The result is that while for the most part the international community does not have the luxury of opting for a strong public agency to control[88] and protect these public goods (the national approach to dealing with public goods), and usually must rely more on co-operation and persuasion, this status takes these public goods a step closer towards receiving the strong unified international attention required to ensure a sustainable climate and to ensure sustainable limits are placed on the emission of ozone damaging substances.

There are instances where the person who receives a commodity is the primary beneficiary and the price that she is willing to pay should, in theory, reflect the benefit, however, other members of society also gain from the provision of the commodity. These are known as impure public goods.[89] For example, energy efficiency grants provide homeowners with the direct benefit of lower heating costs yet society also benefits from a reduction in the consumption of valuable resources such as fossil fuels and a reduction in pollution. Regulation of this sort is clearly directed at preventing harm and clearly is promoting more sustainable energy consumption. Furthermore, while the programme in its initial stages may not be in line with the polluter pays principle, the process may educate homeowners about their resource consumption and encourage some to take greater responsibility for it in the longer term.

---

[84] Birnie, Boyle *op. cit.* pp.390–391
[85] 26 ILM (1987), 1529.
[86] and in UN Framework Convention on Climate Change 31 *ILM* (1992), 851.
[87] Boyle *op. cit.* p.54.
[88] Largely due to states reluctance to give up national sovereignty.
[89] The most common example is free public education.

Information Deficits

For a competitive market to function well, buyers must have sufficient information to identify alternative choices and understand the characteristics of the choices they confront. Furthermore, the market system assumes the choices consumers eventually make are rational. The failure of either of these two assumptions is sufficient to warrant regulatory intervention.

In well functioning markets one would expect to find as much information available as consumers are willing to pay for in order to lower the cost or to improve the quality of their choices. Economists accept that perfect information is a lost ideal and instead prefer to rely on whether the consumer is receiving 'optimal information' where the marginal costs of supplying and processing the level and quality of information in question are approximately equal to the marginal benefits created. For example, does the market encourage car dealers to give consumers enough information about the price, quality and environmental effects of different types of cars to allow them to make a relatively informed decisions? The dealers provide the information that they think will sway consumers towards their product. If the dealers find they sell more cars when they provide the information, they will continue to do so. If the reverse is true, they will likely provide less information. In either case what is provided may not be optimal for the consumer. Unfortunately, the value of information is not known until it is acquired so it is difficult to determine precisely what is optimal. For this reason, policy makers tend to focus on correcting obvious sub-optimal situations.[90]

Information can be deficient or sub-optimal in several ways. First, there may be a lack of information available. People cannot make decisions about whether a product is more or less environmentally friendly if this information is not available to them.[91] Second, the information may be present but inaccurate or incomplete, reflecting only good points about a product. This will obviously skew demand and as a result the market. Furthermore, some information is easier to convey than other information. While it is quite easy to convey information as to price, it is far more difficult to convey information about quality. Regulation is often justified as a way of ensuring consumers do not simply make their decisions based on price. A great deal of environmental regulation is justified the grounds of improving the public's knowledge a particular product or process' effect on the environment: for instance, its recyclability or energy efficiency.[92]

As a result, optimal information is also an important requirement for ensuring environmental justice. The public is not apathetic to the plight of the environment.[93] Improved access to information is vital to ensuring decisions

[90] Ogus *op. cit.* p.39.

[91] See EC Regulation 880/92 [1992] OJ L99/1 which sets up a Community wide system for eco-labelling.

[92] *ibid.*

[93] For example the Royal Society for the Protection of Birds now has over 1 million members in the UK. RSPB 'Saving the Birds, Saving the Future' RSPB Annual Review 1996/1997. See also MORI 'Public attitudes towards nature conservation' a report of a

are made which prevent harm to the environment both now and in the future. It helps individuals fulfil their stewardship role and helps ensure that the polluter pays. If, assuming rational consumer behaviour[94], consumers know a certain product is 'greener' (lithium batteries) than another similar product (nickel cadmium batteries) and then this added information may serve to alter consumer choice in favour of the least harmful alternative.[95]

However, environmental justice also tends to focus on the actual decision making process or on how the regulator should actually regulate a particular problem. Even with 'optimal information' different people will make different choices. Some will focus on price, others on quality. In fact, many people's preferences may not be utility maximising. The best economists can hope for is that individuals act within the limits of 'bounded rationality' accepting that individuals have only a limited capacity to receive, store and process information.[96] This seems to be a safe economic assumption. However, an informed rational decision in the economic sense may not be the same as a decision by a market actor to act responsibly towards the environment. Different decision makers will give different factors more or less weight depending on their own circumstances and values. Some environmentalists would argue that only those decisions which are consistent with the principles of sustainable development and the precautionary principle are rational and utility maximising. While this may arguably be true for the effects of such decisions on society and the ecosystem as a whole, it is not necessarily the case for individuals making individual decisions. For example, in deciding whether or not to recycle, an individual may have optimal information as to where the facilities are located, what can and can not be recycled, the benefits of recycling over disposal etc. However, if the recyclables are not picked up with the rubbish at every home and the individual in question has mobility difficulties then the rational decision may not be the most environmentally friendly decision.[97]

---

[93] (continued) survey for the Nature Conservancy Council 1987; Department of the Environment 1993 Survey of Public Attitudes to the Environment: England and Wales, 1994; Wilkinson, Waterton. Public Attitudes to the Environment in Scotland (Scottish Office Central Research Unit, 1991).

[94] All other things being equal, this consumer would prefer products which protect the environment to those which do not. The assumption being that most people would prefer a clean environment for themselves over a dirty one although they may also favour a clean environment for its own sake and for the sake of others, including future generations. See Sunstein op. cit. pp. 57–59.

[95] The alternative solution is to ban certain products. Regulation which simply alters consumer choice is generally perceived as preferable to bans in a market economy as it interferes less with individual freedom. See paternalism below.

[96] Efficiency theory cannot deal with decision makers who act irrationally. Ogus op. cit. p.41.

[97] See also the decision surrounding the disposal of the Brent Spar oil platform for an example of the difficulties of balancing scientific evidence against social and environmental aims. Kirk "The 1996 Protocol to the London Dumping Convention and the Brent Spar" (1997) ICLQ p. 957

One final point should be highlighted at this stage. Regulation to increase public access to information justified on improving market decisions must be distinguished from regulation which is designed to increase public access to information on the grounds of improving industry accountability, public participation in decision making and adding legitimacy to the regulatory process. Regulation justified on these grounds is discussed below.

Co-ordination Problems

Regulation, not unlike the private law, can be used to facilitate the co-ordination of utility-maximising activities. Tort and property law can act to save transaction costs by setting out standards of behaviour that it is assumed the parties would have agreed to in contracts if transactions costs had not inhibited them.[98] Some co-ordination problems are so complex, that in terms of transaction costs it is more efficient for the public law to provide the appropriate conduct controlling mechanism. This type of regulation tends not to be coercive in nature but rather acts to facilitate market transactions. For example, the protection afford by one state to the feeding grounds of a particular species may be circumvented if its breeding grounds in another state are being systematically destroyed. The European network of protected habitats (Natura 2000) is designed to co-ordinate the legislative action of states to ensure the habitats of migrating and other species across Europe are protected in their entirety.[99] This legislation co-ordinates the actions of states to ensure efforts to prevent harm, take precautionary action and maintain sustainable systems are effective.

Scarcity

Temporary regulation can be justified by exceptional market conditions including acute shortages of supply or uncontrolled demand. The market solution for a shortage of food would be to allocate food on the basis of willingness to pay. The price would rise so much that few would be able to pay. During the first and second World Wars, this concern was used to justify the rationing of food and other necessities. Similarly, the market system affords a nil value to animals and plants in their wild state and they only have value once caught. This leaves wild animals and plants open to exploitation.[100] Recent moves to protect bio-diversity through bans on the killing of endangered species such as whales, elephants and seals can be justified by the fear of uncontrolled demand and shortages of supply. These bans, quotas and rationing which promote bio-diversity do so under the auspices of environmental justice. This action is likely prompted by a concern for the future of the endangered species. It may also be based on a desire to preserve certain creatures for the benefit of future generations of humans. The difficulty lies in determining the appropriate regulatory

---

[98] Ogus *op. cit.* p.41.

[99] EC Directive 92/43 [1992] OJ L206/7 on Habitat and Species Protection.

[100] Some would argue that this problem is actually due to inadequate private rights. If landowners owned the wild animals on their land they would do more to protect them. An example is salmon fishing rights in Britain.

action and taking into account the time lags between the perception of the problem and the adoption of interventionist measures. Indeed attention is now turning to how permanent these bans should be given recent concerns over elephants stomping through established communities and an overpopulation of seals devouring salmon stocks.

Regional Considerations

The market can also aggravate certain undesirable symptoms of the cyclical pattern of demand present in the economy. That is, a gradual increase in the discrepancies between the haves and have nots. In terms of environmental protection, a key macro-economic difficulty to be overcome is that of regional disparity created by the market whereby residents of richer areas are more likely to complain of environmental degradation before those residents in poorer areas. Environmental controls are likely to be tighter in wealthy countries thus, encouraging industry to set up in poorer countries with lax environmental controls. According to conventional theory, the market can be left to solve the problem.[101] As industries relocate to poor countries, this increases wealth creation in those countries which in turn leads to a levelling up of standards and wealth. However, the industrial shifts may never occur as neither labour or capital are completely mobile.[102] Moreover, during the time before levelling up occurs, the damage to the environment may be considered unacceptable and/or permanent: For example substantial numbers of species may become extinct and there may be an increased incidence of pollution related diseases such as asthma and certain cancers. As indicated earlier, the private law may serve only to aggravate these social, economic and ecological discrepancies.[103] This situation conflicts with environmental justice particularly in terms of equity and sustainable development and regulatory action in these circumstances is likely warranted.[104] These issues arose with the shipping of hazardous wastes to underdeveloped countries from the developed countries and they were corrected by international regulatory action.[105]

Non-Economic Goals

In the past, public interest theorists may have ended their list of public interest goals at this point. However, it is now recognised that regulation is introduced for more public interest reasons than those which ensure the efficient running of the market. Four of the key non-economic public interest grounds including: procedural justice, distributional justice, community values and paternalism, will now be examined.

---

[101] See LeGrand, Robinson *The Economic of Social Problems: The Market versus the State* (2nd edn 1984) Ch. 9.

[102] Ogus *op. cit.* p.43.

[103] *Sturges v. Bridgman* (1879) 11 ChD 852: "What would be a nuisance in Belgrave Square would not necessarily be so in Bermondsey".

[104] See Bryant and Grant *op. cit.* above.

[105] Convention on the Control of Transboundary Movements of Hazardous Wastes and their Disposal (Basel) 28 *ILM* (1989) 657 In force 24 May 1992.

Justice

Sometimes the market left unchecked fails to secure a fair process or just distribution. Similarly, government intervention can also result in unjust outcomes. Justice can mean many things to many people.[106] Some argue that distributions are just so long as the process by which resources are acquired is just.[107] At the other extreme, some argue for a more egalitarian approach which requires a redistribution of wealth and power from richer to poorer.[108] The amount of intervention that can be justified in the pursuit of justice is dictated by ideology. Public lawyers tend to distinguish between two types of justice: procedural and distributional. Both types of justice are vital to the goal of environmental justice[109] and will be dealt with in turn, in the context of the market and in the context of government intervention.

Procedural Justice

Even those most committed to a free market system would recognise that procedural fairness is essential in private transactions. The market system, in its ideal state, does promote some aspects of justice. For example, competition ensures consumers are not at the mercy of one unscrupulous supplier. Tort and criminal law protect certain inherent rights such as private property and personal integrity to ensure a fair process. Furthermore, contract law has evolved to ensure market transactions are free from evils such as misrepresentation. However, there are instances where the market assumptions are not met and some regulation of the process is necessary to ensure a fair result. For example, the assumption that the most efficient allocation of resources is achieved by free-market forces rests in part upon an assumption that there is a proper allocation of bargaining power among the parties. Where the existing division of such bargaining power is unequal it may be thought that regulation is justified to provide a better balance. Procedural standards can be used to improve the information available to the weaker party. They may also be used to allow sellers to organise in order to deal more effectively with the buyer. This is particularly the case with labour, fishing and agriculture co-operatives.[110]

Procedural fairness is also needed in the exercise of government power when this is used to correct market deficiencies. For example, since unregulated development of land can result in serious amenity, infrastructure and environmental consequences, most states regulate the use and development of land through planning regimes. When planning permission is refused in the United Kingdom, the applicant must be given reasons for the refusal and is entitled to appeal the decision to a higher authority.[111] In this context, proce-

---

[106] See discussion on environmental justice above at p. and Ogus *op. cit.* p. 46–47.
[107] Nozick, *Anarchy, State and Utopia* (1974) p. 169–72.
[108] For different socialist approaches see George, Wilding, *Ideology and Social Welfare* (rev. edn.) (1985) Ch. 3; Miller, *Market, State and Community: Theoretical Foundations of Market Socialism* (1984).
[109] See p.13 above.
[110] Breyer *op. cit.* p.79.
[111] The Secretary of State: Town and Country Planning Act 1990 s.78(1).

dural standards are commonly thought of as requirements of justice which are needed to establish an appropriate balance in the exercise of governmental power between the citizen and the state.[112] Criteria such as the right to be heard, a right to reasoned decisions and access to the courts to challenge actions and decisions of administrators are regarded as important in the context of achieving this balance.

This approach views the decision-making process as in effect a narrow contest between the individual and the state and is consistent with market assumptions. However, in a society where justice is often associated with individual rights, the environment may have no champion. Environmental resources often have no owner per se. Judicial standing and other forms of recognition have historically been limited to circumstances where a human interest has been affected. Under an ethic based on private property, capitalism and individualism, the environment is bound to suffer.

However, the process is also "a determination of what ought to be done in the public interest in a particular case".[113] Indeed, principles and procedures developed to safeguard the individual against the power of the state are not necessarily sufficient or appropriate for promoting the public interest. The focus for the procedural aspects of environmental justice, as a result, is less on concerns for safeguarding the citizen against the power of the state (although these safeguards are clearly important) and more on a concern for social arrangements for facilitating policy implementation (efficiency), improving accountability and giving the process legitimacy.[114]

These procedural requirements of justice are important both in justifying the type or form of regulation to be used in a particular context and in deciding whether any intervention is warranted or not. One way of improving accountability and efficiency is to tell the public what is going on and then give them opportunities to express their views in the decision-making process.

The benefits of public participation in regulatory decision making are numerous. Some of the reasons, such as improving consumer decision making, have been dealt with above in regard to the correction of information deficits in the market. However, many other benefits relate to the goals of procedural justice. First, public participation reassures, legitimises and promotes public confidence in action taken by Government or industry (secrecy tends to fuel

---

[112] Committee of the Ministers of the Council of Europe, (1977) Resolution (77)31 "On the Protection of the Individual in relation to the Acts of Administrative Authorities"; JUSTICE-All Souls *Administrative Justice: Some Necessary Reforms* (Oxford, Clarendon Press, 1988)

[113] Ganz, *Administrative Procedures* (Sweet and Maxwell, 1974)

[114] Efficiency, in this context, ensures "decisions are taken in a rational and systematic way; that internal conflicts and inconsistencies are brought to the surface and resolved; and that objectives are defined and optimal means secured." Accountability incorporates values such as responsibility, transparency, answerability and responsiveness. Wass 1984 p. 5. See also Harlow, Rawlings *Law and Administration* (London, Butterworths, 1997), Oliver *Government in the United Kingdom: the Search for Accountability, Effectiveness and Citizenship* (Open University Press, 1991).

fear). Second, public participation promotes greater public understanding of the issues involved. Third, public participation can provoke social change and educate the public. Fourth, public involvement can result in more acceptable decisions being taken and less opposition. Finally, public scrutiny can secure more accountability from Government, industry and the voluntary sector.[115] All of these benefits are necessary to secure environmental justice.

There is a growing body of literature that identifies the right of access to environmental information and the right of concerned citizens to participate in decision-making processes as of paramount importance.[116] This is perhaps encapsulated best in Principle 10 of the Rio Declaration which states:

> "Environmental issues are best handled with the participation of all concerned citizens, at the relevant level. At the national level, each individual shall have appropriate access to information concerning the environment that is held by public authorities...and the opportunity to participate in decision-making processes".[117]

The importance of the linkage between mechanisms for public access to information and public participation is also recognised in the Aarhus Convention adopted by the European Environment Ministers in June 1998 and open for ratification by the 53 countries of the UN Commission for Europe.[118] The Convention emphasises the role of certain mechanisms or 'pillars' for public access to information in achieving full public involvement in transparent processes of preparation, monitoring, enforcement and evaluation of environmental decisions.

In the context of the market, the state may have concerns about the perceived accountability of industry to the local community or the legitimacy of the market solution. Regulation (often self-regulation) may be introduced which simply requires industry to monitor its environmental performance and convey this information to the public through an annual report or by way of public meetings. Opportunities such as meetings, lectures, roundtable discussions and public inquiries may also be in place to give the local community some say in how industry operates.

Indeed, there has been a gradual increase in the amount of regulation aimed at improving accountability and legitimacy in both the market and its

---

[115] Rowan-Robinson, Ross, Walton, Rothnie. "Public Access to Environmental Information: A Means to What End?" (1996) *Journal of Environmental Law* 8(1) p. 19.

[116] HMSO 1990 *This Common Inheritance* Cmnd 1200; Boyle, "The Role of International Human Rights Law in the Protection of the Environment" in Boyle, Anderson (eds.) *Human Rights Approaches to Environmental Protection* (Oxford Clarendon Press, 1996); Economic Commission for Europe, Committee on Environmental Policy (1997) Working draft "Elements for the Convention on Access to Environmental Information and Public Participation".

[117] Principle 10, UNECED, United Nations Conference on Environment and Development, *Rio Declaration on Environment and Development.* A/CONF 151/5/Rev. 1.13

[118] UNECE Resolution on Access to Information, Public participation in Decisionmaking and Access to Justice in Environmental Matters, Aarhus, June 1998, ECE/CEP/43/Add.1/Rev.1.

regulatory counterpart. These include regulations which set up public registers, reporting requirements for certain industries, advertising requirements for license applications, public meetings, community forums and, increasingly, the participation of interest groups in the negotiation of voluntary arrangements between the state and industry.[119]

Distributional Justice

In contrast to procedural justice, which focuses on legitimacy and accountability, distributional justice is concerned with the actual distribution of goods and resources between people or groups in society. The determination of what is 'fair' and 'just' is largely a political decision that depends on the particular ideology of the policy maker. For instance, libertarians would argue that so long as the process is just then the distribution is fair. In sharp contrast, socialists seek equality between individuals and groups hence, rely on equality to justify state ownership of the means of production. Liberal theorists tend to take a middle line whereby individuals and groups should have access to a minimum level of resources. This in turn leads to another political debate as to where the line should be drawn in determining 'need'.[120]

Distributional justice is often achieved as an indirect effect of regulation justified by some economic public interest goal. In developing the specific regulatory regime, however, the policy maker may have to determine what is a fair and just outcome. For instance, a choice may have to be made as to whether in correcting an externality, such as pollution, public funds should be used to subsidise new eco-friendly equipment or whether polluters should be taxed on discharges. A subsidy would benefit shareholders and consumers at the expense of the taxpayer while a discharge tax would benefit householders at the expense of shareholders and consumers. The polluter pays principle would dictate that the latter approach be taken while the more sustainable option in the long run may be the subsidy for new equipment.

Regulation can also be justified predominantly on distributional grounds. Ogus argues that redistribution tends to be used to eliminate several types of discrepancy[121]. First, regulation may be used to eliminate or reduce obstacles such as gender, race, disability and wealth which restrict entry into a particular market. Regulation can be also used to eliminate regional discrepancies. For instance, simplified planning zones, enterprise zones and action areas are all designed to encourage development in less fortunate areas of a region or city.[122]

The difficulty lies in determining when is it proper to allow individuals, groups or regions to exploit the advantages which they enjoy in their natural talents (including: intelligence, physical strength and agility, social skills or beauty)

---

[119] See Rehbinder "Ecological Contracts: Agreements between Polluters and Local Communities" in Teubner, Farmer, and Murphy. (eds.) *Environmental law and Ecological Responsibility: the Concept and Practice of Ecological Self-Organisation*, (Wiley, 1994).

[120] Ogus *op. cit.* p. 46.

[121] *ibid.* pp. 48–50.

[122] See for example: Town and Country Planning Act 1990 ss 82–90.

and when redistributive mechanisms should be introduced. Ideological preferences will determine when this is appropriate and where to draw the line.

Redistribution may also be desirable for temporal reasons. The market fails to recognise that the use and consumption of some resources, especially non-renewables, now may vitally affect what is available in the future.[123] The market system fails because the preferences of future generations for relevant resources are not reflected in current demand. Difficulties surround how to predict the preferences of future generations and the alternatives available to them. Again this is an ideological issue. A precautionary approach would dictate that action should be taken now to prevent potential harm[124] from for example, radioactive waste, despite scientific uncertainty over its effects and the capacity of future technology to resolve any problems. Regulation can also be justified to enhance the lives of victims of misfortune where the market through insurance, and the private law through tort have failed. Some risks are simply not insurable. For instance, the cost of a loss in bio-diversity is not insurable nor covered by the private law.

It is important to strike the appropriate balance between allocative efficiency and distributional justice. Too much regulation based on distributive justice will have a negative effect on the workings of the market. Wealth transfers can act as a disincentive for transferees to work and produce. The same effect is felt by transferors who are working to support the transferees. Again the balance is an ideological question. Much of the debate surrounding sustainable development concerns this issue. Sharing technology and wealth can improve our capacity to use resources wisely while still ensuring everyone's needs are met. That said, maintaining or even improving economic and other forms of growth will require incentives to encourage the development of alternative technologies.

Paternalism
Economic efficiency is based on individual choice. Sometimes however, regulatory intervention overrides individual choice. This type of regulation is often justified by 'externalities' resulting in market failure. Paternalistic regulation has to proceed by applying uniform controls on certain activities where it is assumed many individuals make unwise decisions in terms of their own well-being. Sunstein argues that many seemingly unjustified 'paternalistic'

---

[123] This can also be framed as an efficiency problem. Many of the harms environmental laws are designed to prevent are irreversible and certain courses of conduct, such as car dependency, may lead to an outcome from which current and future generations may not be able to recover or may only be able to recover at a very high cost. Since markets only consider the preferences of currant consumers, they do not take account of the effect of transactions on future generations. This results in an externality which may be an irretrievable loss such as the loss of a species or a very expensive back track such as with contaminated land. If future value is to be part of the efficiency calculation, how does one determine it? See Sunstein *op.cit.* p.68.

[124] *Minors Oposa v. Secretary of the Department of Environment and Natural Resources (DENR)* (Supreme Court of the Philippines) 33 *ILM* 173 (1994).

examples of regulation are actually means of facilitating the satisfaction of private desires which may be under attack from social or psychological pressures such as peer pressure. Bans on littering and dog fouling laws provide examples of this type of regulation. Youthful rebellion may actually encourage littering; there may be a stigma attached to picking up after your dog, yet most people recognise the benefits of clean public places. In both of these cases, the law essentially removes or reverses the social pressure and thus helps people do what in fact they want to do.[125]

A lot of paternalistic regulation may also be justified as a means of resolving collective action problems. One person choosing to buy the more expensive refrigerator which does not emit CFCs is not going to make much of a difference to the environment. People need and want a healthy atmosphere yet this is nearly impossible to achieve on an individual basis. Collective action is required. Bans on the manufacture and use of CFCs can be seen as preventative and paternalistic acts by states to protect the ozone layer.[126]

The fact that states took action on the use of CFCs without conclusive scientific evidence means that paternalism can be used in a precautionary sense. However, while the paternalistic solution may be the most sustainable approach to a given environmental problem, it may be inconsistent with notions of shared responsibility and public stewardship. If the paternalistic measure is particularly coercive, it may not even be the most sustainable answer. Indeed, recent moves away from traditional command and control regimes in favour of voluntary and market mechanisms highlight the recognition that these more co-operative approaches often offer more lasting solutions.[127]

Community Values
Some regulation cannot be seen as an attempt to aggregate or trade off private preferences.[128] Instead, it can be justified on the fact that while individuals themselves do not demand a particular good, they believe that the good should be available nonetheless. For instance, someone in Europe still may value the wild beauty of Antarctica or the continued existence of the American bald eagle. Similarly that person may hope that future generations have the opportunity to appreciate drinking straight out of a mountain

[125] Sunstein op. cit. p.50–51.

[126] Convention for the Protection of the Ozone layer (Vienna) UKTS 1 Cm 910, 26 ILM (1987) in force 22 September 1988. See also Montreal Protocol on Substances that Deplete the Ozone Layer UKTS 19 (1990), Cm 977; ILM (1987) 1550 In force 1 January 1989, amended 1990.

[127] Rehbinder op. cit.; Teubner G, Farmer L, and Murphy D. (eds.) Wiley; Ross A., Rowan-Robinson J., (1999) "Behind Closed Doors: The Use of Agreements in the UK to Protect the Environment" Environmental Law Review 1(2) 1999.

[128] For a discussion on the incompatibility between wealth maximisation and non-commodity values (values such as aspirations, mutuality, civic duty) see Stewart "Regulation in a Liberal State: The Role of Non-Commodity Values" (1983) Yale LR. Vol. 92: 1536 at 1563

stream. Sunstein argues that this is because the choices people make as political participants are different from those they make as consumers. As consumers, individuals seek to benefit themselves through their private consumption. In contrast, an individual's political activity is motivated by a variety of factors. They may seek to fulfil individual and collective aspirations through political activity, satisfy altruistic desires which diverge from the self interest of the market; vindicate second order preferences or wishes or even precommit themselves to a course of action in the general interest.[129] Decisions about whether regulatory intervention is justified which reflect the preventative, precautionary principles and / or sustainable development can be based on any of these justifications including concerns for other species, communities and generations.[130]

Conclusions

Given the inherent limitations of the market system in regard to protecting the environment from various harms, considerable reliance is placed on regulation to provide the necessary safeguards. This action is justified as in the 'public interest'. The 'public interest' is defined differently by different individuals, cultures and disciplines. One of the current trends in environmental literature is to define the 'public interest' in environmental concerns as environmental justice. Several environmental principles have emerged as components of environmental justice. The most important of these include: sustainable development, the precautionary principle, the polluter pays principle and the principle of shared responsibility.

In contrast, regulatory theorists tend to link the 'public interest' to those measures required to either ensure efficiency in the market or to achieve some form of distributive justice. Environmental regulation like other forms of social regulation tends to be motivated by three main economic public interest goals: the internalisation of externalities, the provision or protection of public goods and the correction of information deficits. Wider interpretations of 'public interest theory' include non-economic grounds based on distributive justice and altruistic concerns. Public lawyers would probably also add procedural justice to this list.

This paper sought to examine these two sets of public interest goals to discover how they inter-relate. Several conclusions can be made in this regard. First within each discipline the goals are not necessarily consistent. Regulation justified on the basis of making the polluter pay for environmental harm may not in fact be the most sustainable solution. Justice, including environmental justice requires a careful balancing of these conflicting aims. Similarly, regulation aimed at correcting a distributive disparity may produce an inefficient result.

---

[129] Sunstein *op. cit.* p.57.

[130] See for example CITES *op. cit.*; EC Directive 92/43 [1992] OJ L206/7 on Habitat and Species Protection.

Second, the two approaches to 'public interest' differ in one very important respect. Conventional public interest theory tends to be framed in neutral language preferring to avoid the politics behind a given decision to intervene or not to intervene. In contrast, environmental justice and its component principles often are ideologically skewed in favour of environmental protection which, in turn, dictates whether or not intervention is justified. Several further observations arise from this fact.

Most regulation designed to promote environmental justice or more specifically follow some environmental principle can also be justified by one or more of the regulatory public interest objectives. Sometimes, however, the environmental goals will require more, dictating not only whether or not regulation can be justified in the public interest but also the form that regulation should take. For instance, the polluter pays principle agrees that externalities in the market in the form of pollution should be corrected. However, the principle also requires that the polluter should pay to internalise the externality. Thus, while the externality could be corrected either by a subsidy to the polluter for new equipment or by a discharge tax, the polluter pays principle will only be served by the latter.

In other instances, the intervention required to resolve a regulatory problem (such as an information deficit) may be insufficient to meet a related environmental principle. For instance, the precautionary principle not only requires improved public access to information but also relies on the decision-maker (whether they be the policy-maker or the consumer) taking precautionary action towards the environment. This may or may not be consistent with the rational actor model assumed by the market model. Indeed, for a precautionary approach, the decision-maker must receive optimal information and this must be supplemented by regulation justified on altruistic, paternalistic or distributive grounds which force the decision-maker to take a precautionary approach. For example, the state not only informs the consumer about how bad leaded petrol is for the environment but also now prohibits its sale.

While most regulation drafted to fulfil some element of environmental justice can also be justified by one or more of the regulatory public interest goals, the converse conclusion is not true. This is because certain ideological values, which more or less favour environmental protection, are inherent in environmental justice. Public interest theory is not so restricted. For example, regulation aimed at relieving pensioners from the burden of tax on fuel justified on social justice grounds is likely to be inconsistent with the polluter pays principle, the precautionary approach and may even be unsustainable.

Thus, environmental justice objectives can be achieved using the grounds employed by conventional public interest theory. As such, environmental justice appears to be a subset of the broader regulatory goals. This subset does not neatly fall into one of the conventional categories used by public interest theorists. Rather, it is a combination of elements from most of the conventional categories. Thus, there is no need for policy makers to fear justifying environmental protection measures on environmental justice grounds since these are consistent with the traditional justifications aimed at correcting market deficiencies or redistributing wealth. An ideological shift in favour of sustainable development and environmental protection can be accommodated in the same way as socialist, republican or liberal values have been in the past. Indeed, the real hurdle for those promoting environmental justice is that its acceptance requires widespread ideological and cultural change.

Once collective action has been justified, the question then becomes how much regulation is needed and what form it should take. As indicated in the introduction to this volume, the current trend is to develop and experiment with more innovative forms of regulation especially in the areas of environmental protection and exploitation. The goal is to select regulatory mechanisms that can best achieve the collective objective. If the problem is simply a deficiency in the information available then appropriate action may be limited to advertising and reporting obligations. If the problem is one of distributive justice, more intervention may be required. The next two papers in this volume explore the relationship between the objectives of regulatory regimes and the mechanisms used to meet those objectives.

Mechanisms for Environmental Regulation – a Study of Habitat Conservation

Kathryn V. Last

Introduction

Habitat conservation[1] has traditionally relied on voluntary effort, and indeed continues to do so. Since the 1940's, however, the state has assumed increasing responsibility. The approach adopted has been consistent throughout Great Britain up to now,[2] however, as discussed both later in this paper and in further detail in the paper by Colin Reid in this volume, Scottish devolution raises the question of whether this will continue to be the case. This article looks at the purposes of state intervention for habitat conservation and the forms it assumes. It shows how justifications for intervention, under the influence of international and EC obligations, have moved from anthropocentric to ecocentric and shows how initial reliance on direct action and agreements has given way to a much richer variety of forms of intervention.

State intervention for habitat conservation can take a number of forms and this article adopts a four-fold classification of government resources, that can be used individually or in combination, based on Hood's typology: information, wealth, force and direct action.[3] Information is used primarily through consultation between non-departmental public bodies and those whose actions may affect a site protected for its nature conservation interest, for example landowners, local government and the utilities.[4] Wealth is deployed in

---

[1] Habitat conservation must be distinguished from landscape conservation which is effected in England and Wales through designation as a National Park and in Scotland through designation as National Scenic Areas. These areas are primarily protected for their scenic and recreational qualities. Habitat conservation concerns eco-systems protected on scientific grounds.

[2] Unlike landscape conservation which differs greatly.

[3] Referred to as nodality, treasure, authority and organisation by Hood: C. Hood, *The Tools of Government*, Macmillan, London, 1983.

[4] Alternative uses of information are discussed in S. Jacobson *Conserving Wildlife: International Education and Communication Approaches*, Columbia University Press, New York, 1995.

two main forms: the purchase of sites and payments to landowners.[5] Force is utilised via the imposition of criminal sanctions for prohibited conduct[6] and by requirements for ministers, local government and the utilities to consider certain information when making decisions that may affect a site. Direct action is used where non-departmental public bodies carry out restoration of privately owned sites damaged in contravention of prohibitions based on the criminal law or where they control and manage sites. This article focuses on forms of intervention whose primary purpose is habitat conservation.

Recognition of the need to conserve habitats: the establishment of private sanctuaries

Whereas steps had been taken to protect individual animal and bird species by the end of the nineteenth century, state intervention for habitat conservation came relatively late in Britain. The reasons for this were many and varied although it was in part a consequence of the greater emotional neutrality of the subject matter. This was due to the absence of the element of cruelty that was so crucial in providing the impetus for species protection.[7] Rather it was the influence of science, concern about excessive collecting[8] and a change in the concept of the balance of nature that were to prove instrumental in the impetus for habitat conservation.

In particular, the growth of the science of ecology had an important impact: 'At first everything was concentrated on preserving rare or distinctive plants and animals . . . with the development of ecology as a science, the overriding need to protect the habitat of the individual species became more apparent'.[9] The conservation of habitats therefore assumed importance as an essential component of the regime for the protection of species.

The importance of conserving habitats was reinforced by a changing perception of the balance of nature. During the eighteenth century this concept connoted 'a robust, preordained system of checks and balances which

---

[5] Although where purchase of a site is effected through powers of compulsory purchase this will also involve the use of force.

[6] Daintith emphasises that force and criminal law are not the same but that the effect of contravening the criminal law, sanctions, is an example of the use of force. T. Daintith, 'Regulation' in *International Encyclopedia of Comparative Law* Vol. XVII, 1997.

[7] D. Allen, *The Early History of Plant Conservation in Britain*, Proceedings of Leicester Literary and Philosophical Society 34, 1980; B. Harrison, 'Religion and Recreation in Nineteenth Century England' [1967] 38 *Past and Present* 98; and J. Turner, *Reckoning With the Beast: Animals, Pain and Humanity in the Victorian Mind*, John Hopkins Press, London, 1980.

[8] D. Allen, *The Victorian fern Craze: A History of Pteridomania*, Hutchinson, London, 1969. Concern about collecting was also instrumental in the provision of protection for individual species.

[9] J. Sheail, *Nature in Trust: The History of Nature Conservation in Great Britain*, Blackie, London, 1976, p. 196.

ensured permanency and continuity in nature. By the end of the nineteenth century it conveyed the notion of a delicate and intricate equilibrium, easily disrupted and highly sensitive to human interference'.[10] There was a 'reversal of the rationalist, progressivist outlook deriving from the Enlightenment which, with its confidence in the perfectibility of all things, had looked always to the improvement of nature and society through the exercise of human reason'.[11] Actions that had previously been considered as advantageous and for the 'improvement' of nature were now classified as destructive and necessitating control.

Coupled with the industrial revolution, the consequence of this re-evaluation of the concept of nature was that man came to be seen as a major force in nature's destruction.[12] When the rarity of nature and its vulnerability to man's interference was realised, the Victorians and Edwardians adopted a preservationist approach.[13]

The concerns about habitat destruction were, however, essentially anthropocentric. Nature's vulnerability was an issue because of the value of nature to humans.[14] The purpose of habitat conservation was therefore fairly restricted. Conservation was advocated in order to restrict damage to habitats by collecting and thus assist the preservation of individual species considered beneficial to humans.

Because there was no governmental policy on habitat conservation and thus no state intervention in these matters, the task of achieving these objectives was taken on by voluntary organisations concerned with wildlife protection.[15] The prevailing ideology of these groups was that of letting nature alone because of the view that it was direct interference that caused damage. Thus, they established 'sanctuaries', the safety of which lay in the exercise of ordinary property rights which could be used to control access to a site[16] and its management. Any unauthorised interference such as the

---

[10] P. Lowe, 'Values and Institutions in the History of British Nature Conservation' in A. Warren and F. Goldsmith, *Conservation in Perspective*, John Wiley and Sons, London, 1983, p. 337.

[11] P. Lowe and J. Goyder, *Environmental Groups in Politics*, George Allen and Unwin, London, 1983, p. 19.

[12] F. Egerton, 'Changing Concepts of the Balance of Nature' [1973] 48 *Quarterly Review of Biology* 322–350.

[13] This preservationist approach can be contrasted with the US, whose system was based more on an aesthetic interest in nature: A. Warren and F. Goldsmith, *Conservation in Practice*, John Wiley and Sons, London, 1974, chapter 1.

[14] Such anthropocentric concerns were also influential in the earliest protection of wild birds: H. Russell, 'The protection of Wild Birds' [1897] 42 *The Nineteenth Century* 614.

[15] Details of the various organisations formed at this time which were devoted to preserving open spaces and wildlife are given in Lowe, *op. cit.* n.10 at p. 336.

[16] Persons visiting sites can often be the source of damage to those sites. See for example the English Nature press releases: *Guilty – Motorcyclists who Damaged Top Wildlife Site*, 18 November 1999 and *Bank Holiday Rave Destroys Wildlife on National Nature Reserve*, 3 June 1999.

damaging of plants, which are considered to belong to the owner of the soil upon which they are growing,[17] would be actionable at common law.

However, many naturalists were sceptical of the value of such reserves. Because the safety of these reserves lay in the exercise of ordinary property rights, they suffered from a number of limitations.[18] In particular, in Scotland this approach was inappropriate for the protection of wild birds and animals because they are *res nullius* and therefore belong to no one.[19] Furthermore, it was believed that the cost of acquiring and guarding the land would be prohibitive, and that the act of making a reserve would attract the attention of collectors. There was also concern 'at the almost random way in which potential nature reserves were acquired, with apparently little regard for the national significance of their plants and animals'.[20] This was in part due to the organisations responsible for acquiring such sites. By 1910 the National Trust had acquired thirteen sites but 'site selection was haphazard, always secondary to the acquisition of buildings'.[21] This led to the establishment in 1912 of the Society for the Promotion of Nature Reserves whose objectives were 'to preserve for posterity as a national possession some part of our native land, its fauna, flora and geological features'. It concentrated on encouraging other groups to purchase and manage nature reserves rather than doing this themselves.

Nature reserves were therefore considered to be a subsidiary and very expensive means of supplementing legislation to protect species against cruelty and over-collecting; they were merely a stop-gap measure because 'convictions and stringent penalties would soon make watchers redundant and sanctuaries irrelevant to wildlife protection'.[22]

Thus, by the start of the twentieth century there was only a handful of such reserves,[23] and these were largely restricted to England. Habitat conservation was not such a pressing concern in Scotland until sometime later. The problem of collecting was comparatively slight in Scotland and there had been little change in Scottish flora during the preceding century.[24] In fact the National Trust for Scotland was not formed until 1931 some 36 years after its counterpart for England and Wales which was formed in 1895.

---

[17] See for example *Burns v Fleming* (1880) 8 R 226 and *Stewart v Stewart's Exrs* (1761) Mor. 5436 in Scottish law and *Stukeley v Butler* (1615) Hob 168 in English law.

[18] These are well illustrated by the RSPB's first reserve, which had to be abandoned when development on neighbouring land destroyed its natural interest.

[19] D. Carey Miller, *Corporeal Moveables in Scots Law*, W. Green, Edinburgh, 1991, pp. 18–24. Unlike England and Wales where although there is no absolute ownership of wild animals there is a qualified property in them: *Blade v Higgs* (1865) 11 HLCas 621.

[20] Sheail, *op. cit* n.9 at p. 60.

[21] D. Evans, *A History of Nature Conservation in Great Britain*, Routledge, London, 1992, p. 46.

[22] Sheail, *op. cit* n.9 at p. 55.

[23] For details see: Evans, *op. cit* n.21, chapter 3.

[24] *National Parks and the Conservation of Nature in Scotland* (J. D. Ramsey) Cmd 7235, 1947, paragraph 12.

The first state intervention : The National Parks and Access to the Countryside Act 1949

The attitude to sanctuaries in both England and Scotland changed during the inter-war period when changes in land use and management led to the wide-spread destruction of habitats.[25] Urban development and the increased use of land for farming during the war were both factors in the pressure for state inter-vention in habitat conservation. In addition to the nature conservation groups, this pressure also came from groups concerned with amenity and recreation, such as the Council for the Preservation of Rural England established in 1926.[26]

The first governmental response to this pressure was the establishment in 1929 of an inter-departmental committee under the chairmanship of Christopher Addison, to consider the establishment of national parks. The report advocated three objectives for national parks: to safeguard areas of exceptional natural interest against disorderly development and spoilation, to improve the means of access for pedestrians to areas of natural beauty, and to promote measures for the protection of flora and fauna.[27] Habitat conserva-tion thus came to be seen as a subsidiary issue to amenity preservation and recreation. However, the problem that this posed was the irreconcilability of habitat conservation and the provision of recreational facilities. The report gave rise to debate on the value of national parks and habitat conservation.[28] Although the Addison report had considered both of these issues and its remit had covered the whole of Great Britain, the two issues and the situation in Scotland were considered separately in the series of reports that followed.[29]

The case for state intervention was supported by reports from the Society for the Promotion of Nature Reserves and the British Ecological Society. In 1931, the Society for the Promotion of Nature Reserves appointed a Nature Reserves Investigation Committee to develop the case for habitat conservation and to draw up a list of proposed reserves. The subsequent year the British Ecological Society established its own committee to investigate the need for nature

---

[25] Sheail, *op. cit* n.9 at p. 55.

[26] This commonality of interest between the nature conservation, amenity and built heritage preservation groups at this time is discussed in Lowe and Goyder, *op. cit* n.11, chapter 1.

[27] *Report of the National Park Committee* (C. Addison) Cmd 3851, 1931, para. 82.

[28] It is interesting to note that despite the extensive investigation of the issues that fol-lowed, the methods suggested for habitat conservation in the Addison Report closely resemble those ultimately enacted in the National Parks and Access to the Countryside Act 1949: *Ibid.* para. 82(6).

[29] *National Parks in England and Wales* (J. Dower) Cmd 6628, 1945; *Report of the Na-tional Parks Committee (England and Wales)* (A. Hobhouse) Cmd 7121, 1947; *Report of the Wild Life Conservation Special Committee: Conservation of Nature in England and Wales* (J.S. Huxley) Cmd 7122, 1947; *National Parks: a Scottish Survey* (J. D. Ramsey) Cmd 6631, 1945; *National Parks and the Conservation of Nature in Scotland* (J. D. Ramsey) Cmd 7235, 1947; and *Nature Reserves in Scotland* (J. Ritchie) Cmd 7814, 1949.

reserves. The British Ecological Society believed that national parks would be inadequate to meet the requirements of nature conservation: 'In the first place there will be too few of them, and secondly they cannot be managed with primary regard to scientific needs . . . For the purposes of scientific work it is necessary to preserve a considerable number of areas which are generally much smaller, chosen because they represent natural habitats bearing single or several plant communities'.[30] The report also highlighted the risks faced by unprotected sites: 'under existing conditions destructive changes are possible at any time and place as a result of the activities of the speculative builder, of the establishment of new factories or other industrial or public works, of mining and quarrying, and also, though in a different way, of the activity of the Forestry Commission'.[31]

The debate culminated in the formation of two committees specifically concerned with nature conservation: the Wild Life Conservation Special Committee for England and Wales which reported in 1947, the Huxley Report[32] and the Scottish Wild Life Conservation Committee which reported in 1947 and 1949, the Ramsey and Ritchie Reports.[33]

The reports broadly concurred on both the purposes of habitat conservation and the threats faced by habitats. The preservation of amenity, research and education, maintenance of economic value and tourism were the proposed purposes of conservation.[34] The threats to habitats included: development, agricultural expansion, drainage, military use and the taking of particular species for food or collecting.[35] However, the emphasis in the reports differed: the Huxley report focused on the purposes whereas the Ramsey report focused on the threats. Furthermore, the economic purposes were more strongly stressed in the Ramsey report.[36]

Because one of the primary purposes of intervention was the limitation of damage to habitats by development, in finding a suitable mechanism for habitat conservation the emphasis was on land use planning to sort out competing uses. 'In the immediate Post-War period a strong degree of consensus emerged over the appropriate means for protecting the countryside in Britain. Essentially it was assumed that protecting agricultural land from industrial and

---

[30] British Ecological Society, 'Nature Conservation and Nature Reserves' [1944] 32 *Journal of Ecology* 45–82, p. 57.

[31] *Ibid.* p. 49.

[32] *Report of the Wild Life Conservation Special Committee: Conservation of Nature in England and Wales* (J.S. Huxley) Cmd 7122, 1947.

[33] *National Parks and the Conservation of Nature in Scotland* (J. D. Ramsey) Cmd 7235, 1947 Part II and *Nature Reserves in Scotland* (J. Ritchie) Cmd 7814, 1949.

[34] Ramsey *ibid.* paragraphs 5 and 13; and Huxley *op. cit* n.32 paragraphs 31, 35, 39 and 43.

[35] Ramsey *ibid.* paragraphs 7–11; and Huxley *ibid.* paragraph 36. Of the threats identified, agricultural expansion was only mentioned in the Huxley report and the taking of particular species was only mentioned in the Ramsey report.

[36] A similar emphasis on the economic value of conservation in Scotland can be seen in the recent consultation papers: *infra* n.128 & n.129.

residential development and providing an appropriate framework of price support for farmers would combine to produce an attractive rural environment'.[37] This notion of farmers as custodians of the countryside was central to the approach taken[38] with agricultural activities falling outside the controls.

As part of the government approach to habitat conservation, the Nature Conservancy was established in 1949. The primary responsibilities of the Nature Conservancy were to provide scientific advice on the conservation and control of the natural flora and fauna of Great Britain and to establish, maintain and manage nature reserves in Great Britain.[39] These reserves were established under the National Parks and Access to the Countryside Act 1949 which, along with the Town and Country Planning Act 1947 and the Town and Country Planning (Scotland) Act 1947, were the vanguard of state intervention in habitat conservation. The Town and Country Planning Acts concerned development control, the National Parks and Access to the Countryside Act concerned the specifics of which sites of nature conservation interest would benefit from this control and provided additional protection through a range of measures.

The National Parks and Access to the Countryside Act 1949 established a number of different area based designations such as Areas of Outstanding Natural Beauty, National Parks, Nature Reserves and Sites of Special Scientific Interest. Areas of Outstanding Natural Beauty and National Parks, which applied only in England and Wales,[40] were for the preservation of landscape and the provision of recreation and are therefore outside the remit of this article. Nature Reserves and Sites of Special Scientific Interest (SSSIs) were for habitat conservation.

Nature reserves

The Huxley report, whose scheme the National Parks and Access to the Countryside Act 1949 broadly followed, envisaged Nature Reserves to be areas in their natural or nearly natural condition. According to the National Parks and Access to the Countryside Act 1949, they had a dual purpose: the conservation of the area and an educational role. They were areas *managed* 'for the study of,

---

[37] M. Winter, 'Agriculture and Environment: The Integration of Policy?' [1991] *Journal of Law and Society Special Issue* 48.

[38] Huxley *op. cit.* n. 32, paragraphs 133–135 and Society for the Promotion of Nature Reserves, *National Nature Reserves and Conservation Areas in England and Wales: Report by the Nature Reserves Investigation Committee,* Memorandum No. 6, 1945, Proceedings of the Conference on Nature Preservation in Post-War Reconstruction, paragraph 13.

[39] J Sheail, *Nature Conservation in Britain: The Formative Years,* The Stationery Office, London, 1998, p. 33.

[40] For a discussion of why they did not apply to Scotland see: A. MacEwen and M. MacEwen, *Greenprints for the Countryside? The Story of Britain's National Parks,* Allen & Unwin, London, 1987, pp. 10–11.

and research into, matters relating to the fauna and flora of Great Britain and the physical conditions in which they live, and for the study of geological and physiographical features of special interest in the area' and/or for 'preserving flora, fauna or geological or physiographical features of special interest in the area'.[41] Thus direct action and wealth were key mechanisms for the protection of Nature Reserves, combined with a limited use of force.

The Nature Conservancy could buy or lease an area by agreement with the landowner and thereafter would have control over the area and could manage it accordingly, as with the sanctuaries established previously by voluntary organisations. Alternatively, they could enter agreements with the owners and occupiers of land where it appeared to be 'expedient in the national interest' that the land be managed as a Nature Reserve.[42] These agreements could impose restrictions on the exercise of rights over the land or could provide for particular management of the land. Payment for such management or compensation for any restrictions could also be provided in the agreement. An incentive to agreement was provided whereby if the Nature Conservancy was unable to secure an agreement or if breach of an agreement took place, they had the power to compulsorily acquire the land[43] although this was very rarely used.

In addition, the Nature Conservancy was empowered to make bylaws for the protection of nature reserves.[44] However, these bylaws could not restrict public rights of way, and if the site was subject to a nature reserve agreement, the land would remain in its original ownership and the owners' rights would not be affected by these bylaws.

The Nature Reserve designation was limited to a small number of sites because of the criterion of national interest for entering a nature reserve agreement or compulsory purchase and the prohibitive cost of acquiring sites. The SSSI designation therefore provided the protection for the majority of sites,[45] including nature reserves which were also designated as SSSIs.

SSSIs

Unlike Nature Reserves, SSSIs had only one purpose: the conservation of the area.[46] They were seen as 'a means of supplementing with the least interference

---

[41] National Parks and Access to the Countryside Act 1949, s. 15.

[42] *Ibid.* s.16.

[43] *Ibid.* ss. 17 & 18.

[44] *Ibid.* s.20.

[45] By 1981, when the regime was changed, only 171 nature reserves covering 133,640 hectares existed in Great Britain, whereas there were 3,877 SSSIs covering 1,361,404 hectares: Nature Conservancy Council, *8th Report*, 1982.

[46] This is one of the most fundamental differences to the Huxley Report which recommended the introduction of Conservation Areas which had a much greater emphasis on the amenity aspect: Huxley *op. cit.* n.32 paragraphs 56–58. The Ramsey report did not recommend the introduction of such areas in Scotland, only the equivalent of Nature Reserves and Local Reserves: Ramsey *op. cit.* n.33 paragraph 16.

and expense the small range of conditions which have been included in the proposed list of National Reserves' so that it was 'possible to cut to a minimum the sites which should be acquired and placed under strict scientific control'.[47] Direct action was not, therefore part of the protection for SSSIs.

The basis for designation of SSSIs was purely scientific with the selection of sites being carried out objectively on the basis of specified scientific criteria. The statutory criterion for designation was that the area be 'of special interest by reason of any of its flora, fauna, or geological or physiographical features'.[48] The Nature Conservancy was responsible for the selection of sites and it remained for them to decide the conceptual framework and criteria for the determination of special interest. Originally the choice of site was based on the sites proposed in the Huxley and Ritchie Reports but the Nature Conservancy went on to interpret the criteria in a much wider manner.

The reason for the introduction of this designation was to protect sites from damage caused by development. Thus, if the Nature Conservancy considered that a site fulfilled these criteria they were under a duty to notify the planning authority in whose area the land was situated. The purpose of this notification was to enable consultation between the authority and the Nature Conservancy and account to be taken of the special interest of the site when deciding applications for planning permission regarding that site.[49]

This protection for SSSIs was restricted in its application because almost all agricultural and forestry practices were exempt from development control.[50] This was a consequence of the notion of agricultural custodianship and the belief that, 'left alone and protected from urban encroachment, the countryside would take care of itself'.[51] However, this idea was soon to become discredited.

Between 1963 and 1970, the 'Countryside in 1970' conferences were held. These provided important arenas for discussions about the ways in which the countryside had changed since the war and the conflicts that had been generated between the main rural interests. They highlighted the fact that the threats to habitats were changing: 'wildlife was increasingly affected by changes in land use and management, and perhaps most significantly by the ploughing up of old grasslands, drainage schemes, the application of fertilizers and herbicides, and woodland planting programmes'.[52] The National Parks and Access to the Countryside Act 1949 could not provide protection from such damage for the majority of sites as few were designated as nature reserves and SSSIs were not protected from this type of harm.

---

[47] Huxley *op. cit.* n.32 paragraph 58.

[48] National Parks and Access to the Countryside Act 1949, s.23.

[49] Under the Town & Country Planning (Scotland) Act 1947 or the Town & Country Planning Act 1947.

[50] Town and Country Planning (Scotland) Act 1947, s.10(2)(e) and Town and Country Planning Act 1947, s.12(2)(e).

[51] J. Davidson, 'A Changing Countryside' in A. Warren and F. Goldsmith, *Conservation in Practice*, John Wiley and Sons, London, 1974, p. 310.

[52] Sheail, *op. cit.* n.9 at p. 240.

Thus additional protection for SSSIs was provided in the Countryside Act 1968.[53] This gave the Nature Conservancy the power to enter agreements with owners or occupiers of SSSIs. These management agreements were similar to nature reserve agreements and enabled the management of the land in the interests of nature conservation to avoid damaging changes in land use. However, these agreements were rarely used.[54] This may reflect the fact that SSSIs were not regarded as being of such importance as Nature Reserves and the majority of resources were therefore directed at conserving Nature Reserves.[55] Furthermore, although changes in agricultural practice were regarded as a threat to SSSIs, they were not initially treated as seriously as large-scale development because 'many of these changes were piecemeal and aroused little immediate concern'.[56]

In addition, the Nature Conservancy had adopted alternative methods to deal with some of the activities not requiring planning permission. There were arrangements with Agriculture Departments, the Forestry Commission and other public bodies to consult the Nature Conservancy.[57] Such agreements continue to exist, although they now take the form of concordats and statements of intent.[58] This inventiveness on the part of the Nature Conservancy in their dealings with the regime for habitat conservation has continued over the many years of its operation and in part reflects the flexibility inherent in the system. The discretionary nature of the system means that differing political ideologies can be accommodated without a need to change the legislation.

Changing purposes and forms of intervention : The Wildlife and Countryside Act 1981

Views towards agricultural policy began to swing in the late 1970's. Marion Shoard's book *Theft of the Countryside*[59] identified the European Community's Common Agricultural Policy as a key threat to nature conservation. The loss and damage statistics for SSSIs published in 1981[60] showed damage occurring to an estimated 13% of SSSIs in 1980 of which 51% was agricultural in nature.[61] Of particular concern was the fact that 'only a tiny fraction of the

---

[53] Countryside Act 1968, s. 15.
[54] By 1981 there were only 70 such agreements in Great Britain covering only 2,577 hectares, less than 0.2% of the total area designated as SSSIs: Nature Conservancy Council, *8th Report*, 1982.
[55] Nature Conservancy Council, *6th Report*, 1980, p. 14.
[56] Sheail, *op. cit.* n.9 at p. 240.
[57] Nature Conservancy Council, *6th Report*, 1980, p. 12.
[58] For details of those for England see: English Nature, *7th Report*, 1998, p.48 and English Nature, *8th Report*, 1999, p.37. There are none detailed in the reports of Scottish Natural Heritage.
[59] M. Shoard, *The Theft of the Countryside*, Temple Smith, London, 1981.
[60] D. Goode, 'The Threat to Wildlife Habitats' *New Scientist* 22-1-1981, 219–223.
[61] N. Moore, *The Bird of Time*, Cambridge University Press, Cambridge, 1987, p.62.

damage reported was unavoidable'[62] and much of it was not within the ambit
of the existing legislation.

In addition, the implementation of the Directive of the Council of the EEC
on the Conservation of Wild Birds,[63] known as the Birds Directive, required
some changes to British legislation even though it was to a large extent based
on existing British legislation. The Royal Society for the Protection of Birds,
who had been responsible for the drafting of much of the bird protection legislation in the UK, were involved in the drafting of the Directive and the House
of Commons scrutiny committee that considered the proposals also recommended a number of amendments that brought it even closer to the British
legislation.[64]

Although primarily concerned with controlling the hunting and killing of
wild birds and protecting their nests and egges, the Birds Directive required
Member States to take measures to preserve, maintain or re-establish a sufficient diversity and area of habitats for all species of wild birds that occur
naturally in their European territories.[65] The necessary measures included the
creation of protected areas and the upkeep and management of habitats in
accordance with ecological needs.[66] Annex I listed particularly vulnerable species whose habitats were to be subject to special conservation measures. The
most suitable territories were to be classified as Special Protection Areas within
which Member States were required to take steps to avoid pollution or deterioration. Because of the limited number of nature reserves this protection was to
be provided within the UK through designation as SSSIs. However, changes to
the SSSI designation were necessary to ensure compliance with the obligations
under the Directive. In particular, avoiding deterioration of a site[67] was not
achievable with the exiting regime. The need to implement the Directive and
the Ramsar,[68] Berne[69] and Bonn[70] Conventions, led to the enactment of the
Wildlife and Countryside Act 1981.

The regime was, however, very different to that originally proposed by the
government. Radical changes were effected during the passage of the Wildlife
and Countryside Bill through parliament. Originally the government intended
only to introduce what is now known as the Nature Conservation Order to

[62] Nature Conservancy Council, *7th Report*, 1981, p.20.
[63] 79/409/EEC, OJ 1979 L103/1.
[64] For details of the development of the Directive see N. Haigh, *EEC Environmental
Policy and Britain*, Longman, Essex, 1987, chapter 8.
[65] Birds Directive, *op. cit.* n.63, article 3(1).
[66] *Ibid.* article 3(2).
[67] *Ibid.* article 4.
[68] Convention on Wetlands of International Importance Especially as Waterfowl
Habitat 1971 drawn up after a series of international conferences and technical meetings held under the auspices of the International Waterfowl Research Bureau in the
1960s: G. Matthews, *The Ramsar Convention on Wetlands: Its History and Development*, Ramsar Convention Bureau, Gland, 1993.
[69] Convention on the Conservation of European Wildlife and Natural Habitats 1979.
[70] Convention on the Conservation of Migratory Species of Wild Animals 1979.

MECHANISMS FOR ENVIRONMENTAL REGULATION

protect sites covered by international obligations. The basic SSSI designation was to remain untouched with a code of practice being the only change.[71] However, extensive lobbying led to a much wider range of mechanisms being introduced to conserve SSSIs.

The parliamentary debates on the Wildlife and Countryside Bill highlighted the primary purpose of intervention as the minimisation of agricultural damage and in particular the prevention of deliberate harm by landowners. This was a distinct change from the purpose of the National Parks and Access to the Countryside Act 1949. Thus, the emphasis switched from direct action by the Nature Conservancy Council[72] to the use of information and wealth to influence the actions of landowners. This was supplemented with the use of force to try and encourage compromise.

Nature reserves

Because Nature Reserves were either being managed by the Nature Conservancy Council or by the landowner under a nature reserve agreement, agricultural damage did not pose a particular problem. The substantive legislation for nature reserves therefore remained largely the same. The Wildlife and Countryside Act 1981 did, however, provide for a designation known as a National Nature Reserve.[73] National Nature Reserves were nature reserves of national importance that were either owned or occupied by the Nature Conservancy Council, being managed as a Nature Reserve under an agreement, or were owned or occupied by an approved body and being managed as a Nature Reserve. The effect of this was to bring reserves managed by bodies such as the Royal Society for the Protection of Birds within the controls in the National Parks and Access to the Countryside Act 1949 so that bye-laws could be made for such sites.

SSSIs

In contrast, a substantially remodelled form of the SSSI was introduced. This was based on the premise that the use of force should be a last resort and the philosophy of voluntariness should be the predominant characteristic of intervention. Consultation with planning authorities and management agreements were still the basis of protection for sites, however, there were supplemental

---

[71] K. Last, *The Social and Political Determinants in the Formation and Implementation of Habitat Conservation Policy: The Wildlife and Countryside Act 1981* Unpublished PhD Thesis, 1997, Sheffield.

[72] The Nature Conservancy became the Nature Conservancy Council under the Nature Conservancy Council Act 1973. It advised the Government on all aspects of nature conservation and promoted conservation both directly and through the provision of advice and information.

[73] Wildlife and Countryside Act 1981, s.35.

provisions of a mandatory nature which provided for enforced delay in the performance of an operation that may damage a site. This allowed the Nature Conservancy Council time to negotiate a management agreement (with formalised compensation provisions) with the owner or occupier.

Consequently, the provisions focused on the duty to notify the Nature Conservancy Council of proposals to carry out activities on a site. The Nature Conservancy Council could then identify threats to sites and initiate the process of negotiation of a management agreement. The use of force played a very minor role in the scheme, being used to ensure notification to the Nature Conservancy Council of the landowner's intention to carry out a potentially damaging activity on the site.

The criterion for selection of SSSIs set out in the National Parks and Access to the Countryside Act 1949 was replicated in the Wildlife and Countryside Act 1981[74] and if a site fulfils this criterion, the Nature Conservancy Council is under a duty to designate that site.[75] As before, a broad definition has been adopted by the Nature Conservancy Council. In England and Wales there is no right of appeal against designation although in Scotland there is a procedure to refer representations about notification to an advisory committee on SSSIs to consider whether the designation is scientifically justified.[76]

The Nature Conservancy Council is required to notify the Secretary of State, the planning authority and every owner or occupier of the land.[77] All sites notified to the planning authority under the National Parks and Access to the Countryside Act 1949 had to be renotified in this manner if they were to benefit from the additional protection provided by the Wildlife and Countryside Act 1981. Statutory undertakers, the Forestry Commission, the appropriate agriculture department and other relevant bodies are also notified of the designation.[78] The philosophy of voluntariness is highlighted by the fact that a three-month pre-designation period was allowed for the resolution of objections to the proposed designation. However this led to sites being destroyed before designation[79] and this provision was therefore removed in the Wildlife and Countryside (Amendment) Act 1985.[80]

The notification to landowners specifies the reasons for the special interest of the area and any operations that appear to the Nature Conservancy Council to be likely to damage that interest. It is an offence for an owner or occupier[81] to carry out any of these specified operations unless written notice of the

---

[74] *Ibid.* s.28(1).

[75] *R v Nature Conservancy Council, ex parte London Brick Property Limited* [1995] ELM 95, [1996] Env LR 1, [1996] JPL 227.

[76] Natural Heritage (Scotland) Act 1991, s.12.

[77] Wildlife and Countryside Act 1981, s.28(1).

[78] Department of Environment, *Code of Guidance For Sites of Special Scientific Interest*, 1982, paragraph 6.

[79] House of Commons Environment Committee, *Operation and Effectiveness of Part II of the Wildlife and Countryside Act*, Session 1984–1985.

[80] Wildlife and Countryside (Amendment) Act 1985, s.2.

[81] *Southern Water Authority v Nature Conservancy Council* [1992] 3 All ER 481.

intention to carry out the operation has been given to the Nature Conservancy Council[82] and a four-month period has elapsed or the Nature Conservancy Council has consented to the performance of the operation[83] or it is carried out in accordance with a management agreement. There is a defence of 'reasonable excuse' which includes emergency operations or operations authorised by planning permission.[84] The use of force is not, therefore, related to damaging a site but failing to notify the Nature Conservancy Council or failing to wait until the end of the four-month moratorium. Furthermore, the provisions have rarely been enforced[85] due to the unofficial non-prosecution policy of the Nature Conservancy Council.

The four-month period of enforced delay, which can be extended by the agreement of the parties,[86] was designed to allow time for the negotiation of a management agreement. These agreements were originally envisaged a being restrictive in nature, with the landowner agreeing to abstain from the potentially damaging activities which they had notified the Nature Conservancy Council of their intention to undertake. Incentives to enter these agreements were provided in the form of compensation, the amounts of which were laid down in the financial guidelines.[87] The guidelines were not legally binding on the Nature Conservancy Council in all circumstances (only where an operation was notified to them or a farm capital grant had been refused as a result of their objection) but in practice, the guidelines were applied in all circumstances. Payment was calculated in accordance with the principle of full compensation for profit foregone. This included such things as lost revenues had the land been converted to more profitable use and loss of agricultural grants. As such they were generous to landowners; a grant for the work may not have been made in reality but this was of no relevance in determining the amount of compensation.

More recently, the conservation agencies[88] have moved towards a system of agreements that require action by landowners that will positively benefit a site

---

[82] Wildlife and Countryside Act 1981, s.28(5).

[83] It has been estimated that between 75% and 90% of operations are consented to: Livingstone L., Rowan-Robinson J. and Cunningham R., 1990, *Management Agreements for Nature Conservation in Scotland*, Department of Land Economy Occasional Paper, University of Aberdeen, pg. 31.

[84] Wildlife and Countryside Act 1981, s.28(7).

[85] D. Withrington and W. Jones, 'The Enforcement of Conservation Legislation: Protecting Sites of Special Scientific Interest' in W. Howarth and C. Rodgers, *Agriculture Conservation and Land Use: Law and Policy for Rural Areas*, University of Wales Press, Cardiff, 1995.

[86] Wildlife and Countryside Act 1981, s.28(6A). This is an important capability because of the often lengthy negotiation process which is rarely completed within the four month period: Livingstone, Rowan-Robinson and Cunningham *op. cit.* n.83.

[87] Department of the Environment Circular 4/83.

[88] The Environmental Protection Act 1990, s.128, established three separate national bodies to take over the responsibilities of the Nature Conservancy Council. These were the Nature Conservancy Council for England known as English Nature, the Nature

rather than simply refraining from damaging a site. These 'positive' agreements were initially only used in England under the Wildlife Enhancement Scheme introduced by English Nature in 1990. However, they now extend beyond this scheme with the Peatland Management Scheme in the Flow Country of Caithness and Sutherland[89] and a general preference being shown for positive agreements in the management agreement policies of the conservation agencies.[90] Compensation for 'positive' agreements is based on standard payments for particular habitats but with payments tailored to meet the needs of a particular site.

Nature Conservation Orders

A limited number of SSSIs fall within the ambit of Nature Conservation Orders. The Secretary of State or Scottish Minister, in consultation with the conservation agency, may make a Nature Conservation Order if it is considered to be expedient.[91] This order may be applied to a SSSI in two circumstances. The first is for the purpose of securing the survival of any kind of plant or animal, or for compliance with an international obligation.[92] The second is for the purpose of conserving flora, fauna, or geological or physiographical features. The sites must be of special interest and, for those falling under the second category, must be of national importance.

The mechanism is essentially the same as for SSSIs however, the restrictions on carrying out activities are extended to cover all persons and not just landowners[93] and the level of fines is higher.[94] In addition, the four-month negotiation and consultation period can be extended to twelve months.[95] At the end of the twelve-month period, the operation can go ahead without penalty. As with SSSIs, the purpose of the moratorium is to facilitate the formation of a

---

[88] (continued) Conservancy Council for Scotland and the Countryside Council for Wales. In Wales the functions of the Countryside Commission were combined with the Nature Conservancy Council. In Scotland the Natural Heritage (Scotland) Act 1991 merged the Nature Conservancy Council for Scotland with the Countryside Commission for Scotland to form Scottish Natural Heritage.

[89] Scottish Natural Heritage, *Annual Report 1997–1998*, p.6.

[90] English Nature, *7th Report*, 1998, p.29. 'Positive' agreements accounted for 85% of management agreements in England in 1999: English Nature, *8th Report*, 1999, p.36.

[91] Wildlife and Countryside Act 1981, s.29.

[92] For example the Birds Directive, or the Bonn, Berne or Ramsar Conventions.

[93] Wildlife and Countryside Act 1981, s.29(3).

[94] For committing an offence on a site subject to a Nature Conservation Order, the offender is liable to a fine not exceeding the statutory maximum on summary conviction and to an unlimited fine for conviction on indictment. For committing an offence on a SSSI, the offender is liable to a level 4 fine.

[95] This can be done if the conservation agency offer a management agreement or offer to purchase the interest of the person who notified them of their intent to carry out a prohibited activity.

management agreement or the conclusion of terms for purchase of the site. Compensation is payable for the imposition of a nature conservation order[96] and is separate from compensation payable under a management agreement.

An important additional power exists if an offence has been committed on land subject to a Nature Conservation Order. The court has the power to make an order for restoration of the site. This requires the offender to carry out specified works within a specified period to restore the land to its former condition.[97] There is also provision for the conservation agency to perform the restoration and recover any expenses from the landowner.[98]

Nature Conservation Orders impose restrictions for a longer period when SSSI designation has been unsuccessful in securing the protection of the site. If the order also fails, the only recourse is compulsory purchase of the site.[99]

Consultation requirements

The Wildlife and Countryside Act also emphasised intervention through information and a number of consultation requirements were introduced. These were formalised versions of the arrangements made by the Nature Conservancy Council under the National Parks and Access to the Countryside Act 1949.[100]

The conservation agency will inform the utilities of the details of SSSIs within their areas of operation and will ask to be consulted before any activities that are likely to damage the site are performed.[101] This request for consultation is reinforced by a duty to consult when the conservation agency consider land to be of special interest by reason of its flora, fauna or geological or physiographical features and that it may be affected by the schemes, works or other activities carried out or authorised by relevant bodies.[102] In these situations the conservation agency must notify such bodies of the special interest of the land[103] and before carrying out any operations appearing to them[104] to be likely to damage the special features of a site that has been notified, the relevant

---

[96] If there has been any reduction in the value of an agricultural holding due to the imposition of the order: Wildlife and Countryside Act 1981, s.30(2).
[97] *Ibid.* s.31.
[98] Ibid. s.31(6).
[99] National Parks and Access to the Countryside Act 1949, s.17.
[100] *Supra* n. 58.
[101] Department of Environment, *Code of Guidance For Sites of Special Scientific Interest*, 1982, paragraph 11.
[102] Environment Act 1995, s.8. The relevant bodies are the Scottish Environment Protection Agency in Scotland and in England and Wales the Environment Agency, internal drainage boards and water or sewerage undertakers.
[103] Environment Act 1995, ss.8 & 35; Land Drainage Act 1991, s.61C(1); Water Industry Act 1991, s.4(1).
[104] It is the relevant body and not the conservation agency that must consider damage likely.

body must consult the conservation agency.[105] As Ball points out, these duties
are 'ultimately unenforceable, since there is no remedy if the relevant body fails
to consult with the conservation agency. This reflects the fact that their real
purpose is to bring the matter to the NCC's attention so it may give advice or
offer a management agreement'.[106]

The primary consultation obligation continues to be that imposed on planning authorities where planning permission had been applied for on a site
designated as an SSSI. However, this obligation is no longer limited to development proposed on the site itself, because the amended General Development
Orders require consultation for planning applications in any area around a
SSSI defined by the conservation agency as a 'consultation area'.[107] This provision is particularly important for wetland sites where water abstraction some
distance away can cause a lowering of the water table within the site itself. Consultation is to identify operations that may pose a threat to sites that could
either be prevented completely by refusing permission or ameliorated through
the use of planning conditions. Thus planning authorities are required to consult the conservation agency before making a decision and are then subject to
an obligation to take that reply into account when determining applications.[108]
This is supplemented by the Scottish Office National Planning Policy Guide-
line 15 *Natural Heritage* and the Department of the Environment Planning
Policy Guidance 9 *Nature Conservation*.

Implementation of the Habitats Directive : The Conservation (Natural Habitats) Regulations 1994

The EC Directive on the Conservation of Natural Habitat of Wild Flora and
Fauna,[109] known as the Habitats Directive, required further changes to the
scheme of habitat protection in the UK. This was effected through the Conservation (Natural Habitats) Regulations 1994.[110] The aim of the Habitats
Directive is the establishment of a network of Special Areas of Conservation

[105] Environment Act 1995, ss.8(3) & 35(4); Land Drainage Act 1991, s.61C(3); Water
Industry Act 1991, s.4(3). The consultation requirement excludes emergency operations although the conservation agency must be notified of these as soon as possible:
Environment Act 1995, section 8(4); Land Drainage Act 1991, section 61C(4); Water
Industry Act 1991, section 4(4).
[106] S. Ball 'Protected Nature Conservation Sites and the Water Industry' [1990] *Water
Law* 74–78 at 76.
[107] Town and Country Planning (General Development Procedure) (Scotland) Order
1992, and Town and Country Planning (General Development Procedure) Order SI
1995 No 419.
[108] Town and Country Planning (General Development Procedure) (Scotland) Order
1992, article 15 and Town and Country Planning (General Development Procedure)
Order SI 1995 No 419, article 10.
[109] 92/43/EEC, OJ 1992 L206/7.
[110] SI 1994 No 2716.

known as Natura 2000.[111] As with the implementation of the Birds Directive the additional protection was only required for a limited number of sites that are of Community importance.

All sites qualifying as Special Areas of Conservation will already be designated as SSSIs and the Regulations modified the SSSI scheme for these sites in a number of ways. This involved a much more extensive use of force, the most important example being the Special Nature Conservation Order.[112] This differs from a Nature Conservation Order under the Wildlife and Countryside Act 1981 in that the restrictions on damaging operations are permanent. Operations can only be carried out on sites subject to a Special Nature Conservation Order if the conservation agency has given consent or if it is carried out in accordance with the terms of a management agreement.[113] The conservation agencies are subjected to restrictions on the granting of consent for operations[114] and are required to review consents already given.[115] This amounts to an absolute prohibition on certain activities without the agreement of the conservation agency. Furthermore, the regulations empower the conservation agencies to make byelaws for European sites.[116] Despite these changes, management agreements still form the foundation of protection for the majority of sites.[117]

Changes were also made to the application of the development control system to European sites. Under the Wildlife and Countryside Act 1981 the development control system only applied through the consultation requirement placed on planning authorities. Under the Regulations, development that is likely to have a negative effect on a European site can only be given planning permission if there are no alternative solutions and the project must be carried out for imperative reasons of overriding public interest. If the site contains a priority species or habitat, only reasons of human health, public safety and beneficial consequences of primary importance to the environment or reasons which the European Commission consider to be imperative will justify the granting of planning permission.[118] Furthermore, existing planning permissions that are likely to have a significant effect on a European site must be reviewed.[119]

---

[111] This includes sites designated as Special Protection Areas under the Birds Directive. For a fuller discussion of the directive see: P. Birnie 'The European Community and Preservation of Biological Diversity' in M. Bowman and C. Redgwell, *International Law and the Conservation of Biological Diversity*, Kluwer Law International, London, 1996.

[112] Conservation (Natural Habitats) Regulations 1994, reg. 22.

[113] *Ibid.* reg. 23.

[114] If the proposed operation is not directly connected with the management of the site the conservation agency may only consent to it if it will not adversely affect the integrity of the site: *Ibid.* reg. 20 and reg. 24.

[115] *Ibid.* reg. 21(2).

[116] *Ibid.* reg. 28.

[117] *Ibid.* reg. 16.

[118] *Ibid.* reg. 49.

[119] *Ibid.* reg. 50.

The other major change was in relation to permitted development. If such a development is likely to have a significant effect on a European site and is not directly connected with the management of the site, it cannot be commenced or continued until the written approval of the planning authority has been received[120] and the planning authority must consult the conservation agency before giving approval.[121] This provision is particularly important for agricultural activities, which are classed as permitted development, and for development outside a European site that could have an effect within the site.

Problems with the regime for habitat conservation

Although the prevention of agricultural damage was the primary objective of the Wildlife and Countryside Act, the only limitation imposed on these activities was the requirement to notify the conservation agency of the intention to undertake them. This is because most agricultural developments do not require planning permission and therefore fall outside the consultation requirements imposed on planning authorities. Thus the mechanisms introduced in the Wildlife and Countryside Act 1981 for habitat conservation were described as 'toothless'[122] and were subject to heavy criticism by numerous commentators.[123] Although much of this criticism is unfounded because the mechanisms employed have largely achieved their objectives,[124] the regime is failing to protect sites in certain situations, for example neglect and damage by third parties, particularly the utilities, as well as off-site harm.[125] The 1994 Regulations introduced much stricter controls with greater reliance on force that deal with many of these problems. However, this only applies to a limited number of sites. Thus, the majority of SSSIs remain at risk from neglect and damage by third parties. Furthermore, even European sites are not protected from neglect.

---

[120] *Ibid.* reg. 60.

[121] *Ibid.* reg. 62.

[122] Lord Mustill in *Southern Water Authority v Nature Conservancy Council* [1992] 3 All ER 481 at 484.

[123] C. Rose and C. Secret, *Cash or Crisis: The Imminent Failure of the Wildlife and Countryside Act*, British Association of Nature Conservationists and Friends of the Earth, London, 1982; S. Ball, 'Sites of Special Scientific Interest', [1985] *Journal of Planning and Environment Law*, 767–777; W. Adams, 'Sites of Special Scientific Interest and Habitat Protection: Implications of the Wildlife and Countryside Act 1981', [1984] *16(4), Area*, 273–280; Royal Society for the Protection of Birds, *Evidence to the House of Commons Environment Committee*, RSPB Internal Documents, 1984; B. Denyer-Green, 'Why the Wildlife and Countryside Act Has Failed', [1985] *6(1), ECOS*, 9–10.

[124] K. Last 'Habitat Protection: Has the Wildlife and Countryside Act 1981 Made a Difference?' [1999] 11(1) *Journal of Environmental Law* 15–34.

[125] A. Ross and a. Stockdale, 'Multiple Environmental Designations: A Case Study of their Effectiveness for the Ythan Estuary' [1996] 14 *Environment and Planning C: Government and Policy* 89–100.

Proposals for reform

Concern about the implementation of the EC Habitats Directive and the failings of the Wildlife and Countryside Act 1981 in the conservation of SSSIs[126] led to recent proposals to reform the law.[127] As a consequence two consultation papers were published in 1998 prior to Scottish devolution: *People and Nature: A New Approach to SSSI Designations in Scotland* produced by the Scottish Office and *Sites of Special Scientific Interest: Better Protection and Management* produced by the Department of Environment, Transport and the Regions.

Although there are many similarities between the specific proposals in the two consultation papers, they represent different perspectives on the purposes of habitat conservation and the appropriate forms of intervention. The titles alone encapsulate a major difference in attitudes to habitat conservation. The emphasis in the paper for England and Wales is on a continuing move towards an ecocentric approach: "the conservation of SSSIs is not sufficient on its own to deliver the protection of our wildlife but their safeguarding is an essential precondition for the success of that objective".[128] In contrast, the paper for Scotland takes a more anthropocentric stance: "it is essential that proposals to enhance nature conservation in Scotland should be set within the wider framework of sustainable development in rural and other parts of Scotland, having regard also to the need for land reform"[129] and "measures for conserving the natural heritage should wherever possible assist, not impede, the sustainable use of the countryside and its natural resources, taking full account of the social and economic needs of local communities".[130]

The core of the proposals in the Scottish Office consultation paper is the restriction of SSSI designation. The idea is to have a suite of sites that represent 'the jewel in Scotland's nature conservation crown'. Sites that are not considered to be of national importance would be denotifed and responsibility for these sites would be shifted from national to local level. This proposal has important consequences because sites may be important in the context of habitat conservation across the UK but less important in the context of Scottish habitat conservation, for example if such sites are more common in Scotland. Any limitations on the number of sites designated in Scotland could therefore have important ramifications. Furthermore, all SSSIs should be of national importance anyway. The reason for the Scottish report advocating this change may lie in the different nature of designated sites in Scotland. Generally they tend to be much larger than in England and Wales and the perceived "burden"

---

[126] House of Lords Committee on the European Communities, 18th report, 1999.

[127] *Meacher Sets Out Actions to Protect Our Finest Natural Habitats* Department of Environment, Transport and the Regions press release, 2 August 1999.

[128] Department of Environment, Transport and the Regions *Sites of Special Scientific Interest: Better Protection and Management*, 1998, p.5.

[129] Scottish office *People and Nature: A New Approach to SSSI Designations in Scotland*, 1998, p.4.

[130] *Ibid.* p.7.

of designation is therefore more widespread leading to greater pressure to reduce the number of designated areas.

Another central concept in the Scottish Office paper is the involvement of the local community. The Scottish view seems to be that better protection will be afforded to sites by raising public awareness and involvement. The belief posited is that more involvement by local communities, landowners and occupiers will lead to greater commitment by these people to protect sites, particularly if the sites can be shown to be economically advantageous to the area. Thus, a proposed pre-notification consultation process would include the local authority and local groups as well as owners and occupiers. It is also proposed that the management of sites and decisions on whether to allow restricted operations to be carried out should take account of the views of the local community and that a right of appeal should be available against the refusal of consent to carry out a restricted operation which would also take account of the views of the local authority and local groups. Rights of appeal are also proposed for challenging designation and the inclusion of a particular operation on the list of restricted operations. The English paper did not dismiss local involvement but seemed to regard it as a peripheral consideration.

The primary objectives for reform of habitat conservation in Scotland therefore seem to be the limiting of the SSSI network and the transfer of much responsibility away from Scottish Natural Heritage to a local level. The secondary objective, which is the primary objective in England, is the strengthening of the existing controls for SSSIs to close the gaps that exist in the current protective regime. This different emphasis has consequences for the mechanisms likely to be favoured in the implementation of these objectives.

The political climate has changed since the publication of these two consultation papers and responsibility for legislating on environmental protection and nature conservation has been devolved to the Scottish Parliament.[131] However, responsibility for compliance with the EC Habitats Directive lies with Westminster, thus the approach to nature conservation may be flawed if it is not UK-wide.[132]

The Countryside and Rights of Way Bill

The favoured mechanisms in England and Wales are now set out in the Countryside and Rights of Way Bill,[133] presented to Parliament in March 2000. This involves much stricter controls for SSSIs, with a greater reliance on force. The most fundamental change is the permanent restriction on potentially damaging operations. These can only be carried out with the consent of the conservation agency, in accordance with a management agreement or in accordance with a 'management scheme' or 'management notice'.[134] They can no

---

[131] Scotland Act 1998, Schedule 5.

[132] Ross and Stockdale, *op. cit.* n.125.

[133] Bill 78 – EN.

[134] Countryside and Rights of Way Bill, Schedule 9 paragraph 1, clause 28C(3).

longer go ahead after a four-month period. However, if consent has been refused for an operation there is a right of appeal.[135] This scheme would bring protection for SSSIs in line with that for European sites and through the management scheme and management notice it would deal with the defects of the existing regime in dealing with neglect.

A management scheme for conserving or restoring the features of interest on a site is formulated by the conservation agency after consultation with the owners and occupiers of the land.[136] After notice of the proposed scheme has been served on the owners and occupiers they may make representations. This provision enables the conservation agency to determine the management of a site without an agreement with the owners and occupiers but payment can be made.[137] If a management scheme is not being implemented and the special features of the site are being inadequately conserved or restored as a consequence, the conservation agency can issue a management notice if they are unable to conclude a management agreement.[138] A management notice can require specific works that are reasonably necessary to manage the land in accordance with the management scheme. If these are not carried out, an offence has been committed[139] and the conservation agency can carry out the work and recover the costs from the owner or occupier.[140] Similarly, the power to order restoration of a site is extended to all SSSIs.[141]

The power of compulsory purchase is also extended to all sites[142] and the levels of fines are greatly increased.[143] Furthermore, third parties intentionally or recklessly destroying or damaging the special features of a site will be guilty of an offence if they knew that it was within a SSSI.[144]

Government departments, local authorities and statutory undertakers are subject to a requirement to notify the conservation agency before carrying out operations likely to damage the special interest of a SSSI. If the conservation agency refuses to assent to the operation it may be carried out if 28 days notice has been given to the conservation agency stating how account has been taken of any advice given by the conservation agency and it is carried out in such a way as to result in as little damage as is reasonably practicable.[145]

Because of the much stricter controls imposed by this scheme, the Bill provides for the repeal of s.29 of the Wildlife and Countryside Act 1981 (which

---

[135] *Ibid.* clause 28D.

[136] *Ibid.* clause 28H.

[137] *Ibid.* clause 28K(2).

[138] *Ibid.* clause 28I.

[139] *Ibid.* clause 28M(6).

[140] *Ibid..* clause 28I(7).

[141] *Ibid.* clause 28(3).

[142] This may be done where the conservation agency is unable to conclude a management agreement or one has been entered into but has been breached: *Ibid.* clause 28L.

[143] On summary conviction to a fine not exceeding £20,000: *Ibid.* clause 28M(1). Cf. The current levels of fine *supra* n.94.

[144] *Ibid.* clause 28M(5).

[145] *Ibid.* clause 28F.

gave the conservation agencies the power to make Nature Conservation Orders). Because damaging operations can be subject to permanent restriction, the nature conservation order is no longer necessary.

The provisions in the Countryside and Rights of Way Bill match very closely with the proposals in the consultation paper. It is interesting to note that the proposal to extend the restrictions on activities on SSSIs beyond a four-month period and making the moratorium on operations permanent was never referred to explicitly in the Scottish report[146] although it was in the report for England and Wales. It is therefore uncertain whether similar provisions will be proposed by the Scottish Parliament even though the proposals regarding the strengthening of controls are very similar in both papers.

For example, one proposal for consideration in the Scottish paper was the introduction of an order-making power, requiring the owner to manage the land in a particular way, with default powers so that Scottish Natural Heritage could enter, carry out the work and recover the costs. This is effectively the management scheme and management notice procedure provided in the English Bill. However, the Scottish paper also proposed limiting the use of force in some circumstances by dispensing with restrictions on the activities that can be undertaken on sites where the owner has agreed a 'conservation contract' or where there are considered to be other avenues for protection of the site.[147]

Both papers advocated changes in the payments associated with management agreements. Thus, it was proposed that the presumption in respect of compensation for abstaining from damaging activities be removed in England and be limited to exceptional circumstances in Scotland. This has been achieved in the English proposals by extending the restrictions on operations. Thus, the Countryside and Rights of Way Bill will formally shift payments from compensation for abstention to those under positive management agreements or management schemes, reflecting the position that has almost been achieved by the changing policies of the conservation agencies on management agreements.

One of the main differences between the two papers was the emphasis in the Scottish paper on community involvement. Although community involvement may be a laudable aim it is not clear what real benefit will be obtained by the proposals for such input into the designation and management of sites where the fundamental issue is scientific and where the expertise is unlikely to be available in the community. The reliance on other indirect protection mechanisms, such as agricultural grants and development control, also faces the problem of lack of expertise and may provide less protection than currently exists. Many of the proposals may therefore be seen as a retrograde step, devaluing the importance of nature conservation and limiting the effectiveness of the SSSI designation in Scotland.

---

[146] However, it is strongly hinted at in some of the other proposals.

[147] For example, where the operations which would be likely to have a significant effect on a SSSI were subject to a statutory consent by another body or dependent on public grant or subsidy.

Conclusions

The objectives for habitat conservation have evolved greatly since nature reserves were first advocated. Although the objectives were initially anthropocentric and rather general, protecting sites from human interference, they have become more ecocentric and more specific about the particular sources of harm to be prevented. The mechanisms have also evolved to take account of these changing objectives. As the objectives have become more specific, reliance on force has increased and deference to private property rights has diminished. However, there has also been a move towards greater 'environmental justice'. In particular, the Countryside and Rights of Way Bill shows a much greater emphasis on the precautionary approach and to some degree the polluter pays principle.

Although the Scottish consultation paper also displays such trends it raises potential problems for habitat conservation in Scotland. Given the desire to co-ordinate any proposals for habitat conservation with land reform proposals, differences in the schemes for habitat conservation are highly likely. This presents a particular problem for cross-border sites. Furthermore, dealing with some of the gaps in protection, such as damage by the utilities, may be practically difficult because of a lack of competence in the Scottish Parliament. That said, it is not clear that the Scottish Executive intends to follow the proposals set out in the consultation paper issued by the Scottish Office. Scottish Natural Heritage are currently debating the point with the Scottish Executive and any legislation may differ significantly from what was proposed prior to devolution.

Improving Land Use Management Through Economic Instruments[1]

Nick Hanley and Douglas MacMillan

Introduction

Environmental problems are often attributed to the impersonal face of market forces. For instance, the expansion of private sector coniferous afforestation in Scotland in the 1980s in sensitive areas such as the Flow Country was attributed to the availability of tax concessions and planting grants which made planting an attractive proposition. Similarly, the great expansion of agricultural activity in the 1970s and 80s and the environmental problems which followed from this are often attributed to the more favourable terms available to farmers once Britain joined the EEC. Economists would describe these types of incidents as *policy failures*, since they resulted from government intervention in the market which had undesirable environmental consequences. Economists also point out, however, that the market itself it often unkind to the environment, since environmental resources are under-valued due to a lack of clearly defined property rights. The fact that firms face no market charge for polluting the atmosphere means they have little market-derived incentive to cut back on emissions. The fact that farmers receive no payment from the market for producing beautiful landscapes means that too few are produced in a pure market system. Both of these effects are referred to as *market failure.*

Governments have thus long intervened in the market on environmental grounds, seeking to produce a higher level of environmental quality than the market would produce. These environmental gains, be they in terms of lower river pollution, cleaner air or more habitat protection, are important from an economic point of view, since they constitute real economic benefits (Hanley and Spash, 1993). Thus, even though a cleaner river generates few benefits valued by the market, this environmental improvement has economic value if it impacts positively either on people's utility (a jargon term for satisfaction or happiness), or on the production activities of other firms (eg fishermen).

---

[1] Thanks are due to Andrea Ross-Robertson for helpful comments on an earlier draft of this paper.

There are many ways in which governments can intervene to produce a greater level of environmental quality than the free market system would generate. Consider the example of pollution control. Regulation is one means of control open to a government agency, either in terms of insisting on reductions in polluting emissions, or in the adaptation of cleaner technology. These requirements are then backed up by legal sanction. Alternatively, the government can use persuasion, to encourage firms to adopt cleaner production methods, perhaps in order to achieve market-valued advantages from green credentials. Another approach would be to make use of the concept of legal liability, for instance through making firms liable for environmental damages associated with their actions. For example, the Superfund legislation in the USA makes certain firms legally responsible for environmental damages resulting from their activities, with injured parties being able to sue for monetary compensation. Finally, the agency could make use of *economic instruments*. But what are these, and how do they work? We explain using the example of rural land management.

Economic instruments are policy interventions that change the financial incentives that land owners or managers face, in such a way that behaviour changes to bring us closer to some social goal. By paying people or firms to produce environmental goods, more will get produced. Agri-environmental schemes operate in exactly this way, by offering a financial incentive for farmers to "produce" better wildlife habitats on their farms, for example by restoring wetlands or extending farm woodlands. Equivalently, by increasing the cost of something, we can encourage people to use less of it. Taxing pesticides will encourage farmers to use less of these products, or if highest tax rates are put on most environmentally-damaging substances, to switch to less damaging substances. Placing an environmental tax on nitrate fertilisers or livestock manure applications in catchments subject to eutrophication will encourage farmers to apply less of these potentially-polluting inputs to land. In a broader context, putting an environmental tax on land-filling should encourage people to throw away less and recycle or re-use more.

Economic instruments may also be used to create tradeable entitlements in resources whose ownership was previously unclear or undefined. The effect of this is equivalent to putting on price on such resources. For example, concerns are growing regarding abstraction of water from rivers, and associated environmental damages during low-flow periods. In Scotland, abstraction rights are poorly defined. By creating legal rights and allowing them to be traded, we both control the total amount of abstraction allowed, and put a money value on the right to extract some specified quantity of water. Since water rights are now valuable, a financial incentive is created for users (ie farmers or water companies) to utilise water more efficiently, since the decision to use more water for irrigation, for instance, means fewer rights are available to sell to another user. A similar example is the creation of tradeable discharge rights to control point-source pollution for example from substances which exert a Biological Oxygen Demand. Again, two things happen: the total amount of discharge is controlled, and a financial value is attached to the right to discharge which results in a more efficient use of the assimilative capacity of a river or estuary (Shortle et al, 1999). A final example concerns the creation of tradeable carbon credits in order to meet the UK government's commitment to the Kyoto protocol (see below). A key feature of all of these examples is that the government fixes the level of environmental damage, whilst at the same time making the right to use the environment in some way, valuable.

Why might economic instruments be a good idea?

Just because economic instruments can help us achieve environmental objectives does not mean we should use them, since we have already seen that alternative mechanisms are available, namely regulation, legal liability and voluntary persuasion. For economic instruments to represent a desirable option, they have to be better in some respect(s) than these alternatives. Voluntary approaches which are not backed up with financial incentives or the threat of future regulatory actions are unlikely to be effective if avoiding environmental damages or creating environmental benefits is costly. We thus ignore such voluntary approaches from now on. Liability offers a realistic way forward in relatively few cases (due to the high transactions costs of resolving disputes, and asymmetric information) so we disregard this too. The main comparison is thus between economic incentives and regulation.

Economists have identified a number of important advantages for economic instruments, dating back to early work by Baumol and Oates (1971, 1975). These are as follows:

- Economic instruments allow us to change the level of environmental damages or benefits in a way which permits more flexibility in response than alternatives such as regulation;
- An important corollary of this is that we can achieve an environmental target *more cheaply or efficiently* with an economic instrument than with regulation.
- Economic incentives produce a dynamic, on-going incentive for producers generally to adopt cleaner technology, compared to regulatory alternatives.
- In some cases, economic incentives implement the "polluter pays principle" (for environmental damages) or the "provider gets principle" (for environmental benefits).

The second bullet point is the most important. By allowing flexibility, economic instruments allow us to achieve environmental objectives at a lower cost to society than regulatory alternatives. This can be proved mathematically for a wide range of cases where the cost of achieving environmental objectives varies across agents (Baumol and Oates, 1988). An alternative way of thinking about this property is to say that, for a given total expenditure, economic instruments allow a greater level of environmental improvement than regulatory alternatives. Much empirical evidence now exists to support this finding in the context of industrial pollution control, both in terms of experimental work and from actual policy, in for example the US Sulphur Trading Programme (Sorell and Skea, 1999). Evidence of the cost-effectiveness of economic instruments in the context of land use is also accumulating: see for example Shortle, Abler and Horan (2000) on non-point pollution from farming, and Hanley et al (1998) on heather moorland conservation in Scotland.

However, it should be noted that economic instruments, although increasing in use throughout the OECD, are still used less than the strengths of these arguments might suggest. Reasons why are investigated in Hanley, Hallett and Moffatt (1990), and include political opposition to pollution taxes, practical difficulties in calculating and enforcing emission taxes, and ecological complexity. It should also be pointed out that many environmental taxes are introduced mainly for revenue-raising reasons, rather than as a means of changing behaviour (OCED, 1989).

Leading on from this, we can also suggest some general circumstances where economic instruments are most likely to run into problems. These include:

(a) if an absolute prohibition of actions is required, for example if we are dealing with a highly dangerous substance;
(b) if environmental damages are most likely to result from accidents or other unpredictable events;
(c) if there is great uncertainty about the cost of the environmental damage.

Finally, we note that if we are most worried about achieving targets of environmental quality, rather than about minimising the costs of control then where there is considerable uncertainty over pollution control costs, environmental taxes will perform worse than tradeable entitlements.

We now look at four policy areas, where economic incentives could be used to better achieve environmental objectives relating to rural land use. These are:

- Multi-purpose forests;
- Indigenous species: red deer and wild salmon;
- Managing congestion and erosion at outdoor recreation sites; and
- Agri-environmental policy.

In each of these cases, it is possible to identify uses of economic instruments which avoid the situations referred to in the previous paragraph.

Multi-use forests

Forests generate many benefits for society other than just marketable products like timber. These non-timber values include those associated with forestry's role as a habitat for wildlife, its value as a landscape feature, the role of forests as a location for informal recreation, and carbon sequestration values. These benefits have now been extensively studied in the UK, and economic estimates placed on them using techniques such as contingent valuation and the travel cost method. For example, Hanley (1989) estimated the economic benefits of the Queen Elizabeth Forest Park for recreation and as a wildlife habitat. MacMillan (1990) used estimates of recreation and carbon storage values for a range of forests across Scotland to re-calculate social rates of return. Hanley and Ruffell (1993) and Hanley, Wright and Adamowicz (1998) used contingent valuation, travel costs and choice experiments to estimate landscape values for a range of Forestry Commission sites; whilst MacMillan and Duff (1998) report results on the non-market benefits of restoring native woodland at two sites in Affric and Strathspey. The growth in estimates of non-market values for forests has allowed the Forestry Commission to make increasing use of these estimates in their decision-making.

However, these non-timber benefits are un-priced by the market, so that forest owners do not get rewarded for producing them. Whilst managers of public forests can and do take such benefits into account in their management decisions, this market failure has significant implications for private landowners. The result is too few forests, and the wrong kind of forests. One solution is to make use of economic instruments. For example, differential rates of grant are now offered for the establishment/re-generation of native woodlands, relative to commercial conifer species. The high rate of grant available for native woodlands (£1050/ha) compared to only £700/ha for conifers can be considered to be a greater incentive to produce forests with higher recreation and conservation benefits. The Woodland Grant Scheme is also spatially targeted, with higher rates of grant available for woodlands close to towns where there is considerably greater demand for recreation. As observed by Reid (1995), these changes to the grant scheme have made substantial improvements to the effect of forestry on the countryside. There is more planting on lower ground, more use of broad-leaved species and more sympathetic felling techniques. However, there exists further potential to exploit spatial targeting in forestry policy. For example, Macmillan et al., (1998) have shown how the environmental benefits of native woodland restoration schemes could be enhanced by a carefully differentiated grant design using spatial characteristics. Carbon credits paid to forest owners by the market for the rights over carbon sequestered in these forests, are also an example of an economic instrument which helps correct market failure. Such credits would be a major boost to the Scottish forestry industry, and have been cautiously welcomed in a recent government statement (annex one). Such caution is to be expected since awkward issues of the validation of credits, their lifespan and the net impacts on carbon lock-up (given that ploughing land releases carbon) remain to be resolved.

Indigenous species: red deer and salmon

These species offer an interesting contrast, in that we seem to have too many of the former and insufficient of the latter. How can economic instruments be used to rectify these problems?

Red deer

It is widely acknowledged that the current population of red deer in Scotland is too high in terms of the balance between population demands and ecological carrying capacity. This imposes external costs on other land users, in terms of damages to forest stocks and other crops; and in terms of ecological damage from over-grazing (Hanley and Sumner, 1995). However, the fact that estate capital values are linked to the number of shootable stags present, and that the costs of culling hinds exceeds the revenue so gained, means estates keep stocks at too high a level: for example, in 1995/6, the hind cull was 2000 animals, significantly less than the 27000 needed to prevent the population from rising (Gordon Duff Pennington, 1997).

The emergence of deer management groups may be seen as one privately-led initiative to internalise some of the external costs of red deer, and produce more co-operative outcomes (Hanley and Sumner, 1995; Bullock, 1999). Deer management groups operate as semi-formal collaborative agreements between neighbouring deer estates over issues such as culling rates and access routes during the stalking season. However, it is unlikely that these will successfully address the problem of ecological over-grazing, since the costs fall only partly on estate owners. How could economic instruments help? Three possibilities may be identified:

(i) a subsidy scheme for culling, regionally and even locally varying according to the expected environmental gains from reduced grazing pressure. These would vary spatially, and could be perhaps approximated by indicators of forest re-generation potential.

(ii) a tax-subsidy scheme. This would aim to be self-financing. Each area would be set reference levels for deer population. Estates who brought deer numbers down below this reference level would receive a subsidy payment per deer counted below the level. Each estate with deer numbers in excess of this level would pay a "deer tax" per head of deer above the level.

(iii) a tradeable permit scheme. This could be organised in terms of either tradeable entitlements to keep deer; or in terms of tradeable cull requirements. So long as the costs of culling vary across estates, then such a scheme could reduce deer numbers more efficiently than uniform population reduction requirements (for details, see MacMillan, 2000).

Wild Salmon

Salmon stocks in Scottish rivers have been declining, particularly on the West coast (Watt, Bartels and Barnes, 1999). Economic mechanisms have already been used by rod fishery owners, who have bought up netting rights up-stream, and not used these rights. This has increased the number of fish passing up-river to spawn. However, bankside erosion, fish-farming activities and pollution all act negatively on salmon populations (WWF, 1995). Economic incentives could be used to improve bankside habitat management, for example by making use of agri-environmental scheme payments to farmers to create margins around fields which border on rivers. The environmental implications of fish-farming on wild salmon stocks could perhaps be addressed by an environmental tax being levied on fish farms, with the proceeds being used to regenerate wild salmon populations.

Reducing congestion and erosion in outdoor recreation areas

Although less of a problem in Scotland than in some parts of England such as
the Lake District, concentrations of recreational users can impose externalities
in terms of environmental damage (eg erosion and disruption of sensitive wildlife sites) and in terms of dis-amenity effects on other users (crowding)
(Wightman, 1996). The prognosis is that this will get worse, since the demand
for outdoor recreation is rising faster than the rate of growth of the economy.

Consider for example, erosion and congestion problems at a popular mountain site. One way of addressing this problem with economic incentives is
through pricing. There are two alternatives:

(i)   Access fees, for example through car parking fees. Charging for access to
      mountain areas in other ways (for instance, in a similar manner to US na-
      tional parks) would be deemed impractical and highly unpopular. Car
      parking fees are perhaps more practical in some locations, and have a simi-
      lar impact, but may still be viewed as unpopular and unfair to poorer
      households.
(ii) Iincreases in the time cost of access. Time is a scarce resource for everyone,
      especially leisure time! By moving access points further away, or by ban-
      ning certain means of transport to sites (eg banning cars along private
      estate roads, or banning mountain bikes on estate tracks), it is possible to
      increase the time price of accessing a site. In fact, the "long walk in policy"
      is becoming more widely used in Scotland (for example: the banning of
      mountain bike access to the Southern Cairngorms through the Mar Lodge
      estate).

Of course, as we increase the time or money costs of acessing given sites, we
could expect pressure to increase at other, substitute sites. Hanley et al (2000)
have modelled this process using a data set of rock-climbers in Scotland. Table
1 gives some illustrative results, and shows the effects of car parking and time
price increases both in terms of utility per visit and the number of visits.

Another issue here is cross-activity externalities in a recreation context. For
instance, noisy water skiers at Loch Lomond reduce the pleasure of those preferring quieter activities at the site, such as fishing or walking. Should we
impose a "noise levy" on skiers? This could perhaps vary according to boat
engine size, and would serve as a dis-incentive to those who value a day's water
ski-ing least to change to some quieter sport, or alternative location.

Agri-environmental policy

As we have already argued, farmers can generate both positive and negative
environmental effects through their actions. Positive actions include management which conserves semi-natural habitats such as heather moorland,
haymeadows and heathland. Negative impacts are associated with management decisions which degrade or reduce such habitats, and with non-point

Table 1   Impacts of increasing time or money costs of access in three mountain areas.

Policy option/site	Reduction in seasonal visits per climber, dV	Aggregate dV	Aggregate welfare loss, £/season	Welfare loss per reduced visit, £
A: Ben Nevis: car parking fee of £5	1.3	16,152	161,525	10.00
B: Ben Nevis: time price increase of 2 hours/day	1.85	22,986	155,312	6.76
C: Glencoe: time price increase of 2 hours/day	3.42	42,493	298,200	7.02
D: Cairngorm: car parking fee of £5	2.47	30,689	248,500	8.09
E: Cairngorm: time price increase of 2 hours/day	3.49	43,363	285,775	6.59

Source: Hanley, N, Alvarez-Farizo, B and Shaw, D (2000) "Rationing an open-access resource: mountaineering in Scotland". Mimeo, IERM, University of Edinburgh.

source pollution from the use of fertilizers, manures and pesticides. Again, as mentioned already, the market provides an insufficient incentive from society's point of view for farmers to generate the "best" level of environmental impact.

Agri-environmental policy (AEP) has become an increasingly important component of rural policy in Scotland as in the rest of the UK, and recent government announcements suggest that this trend will continue. The basic model is for farmers to be offered payments in return for management agreements which promise actions designed to increase the production of environmental goods, such as the semi-natural habitats referred to above. Take-up is voluntary. Cumulative spending on initiatives such as the Environmentally Sensitive Areas scheme, the Habitats scheme and the Countryside Premium scheme was £86 million in 1997 (Hanley, Whitby and Simpson, 1999).

The overall evidence suggests that this use of economic instruments to produce environmental benefits from farm land has been highly efficient, in the sense that benefits outweigh costs by a very considerable amount. This also suggests that increasing expenditure on such schemes would be desirable (Hanley , Whitby and Simpson, 1999), although we have few estimates of

marginal benefits. However, it is also apparent that there is scope for considerable improvements in how schemes are designed, since for example the use of
fixed-rate payment schemes means some farmers get over-compensated for
the costs of participating, whilst the overall costs of the scheme could be reduced by allowing farmers to bid for conservation contracts (Latacz-Lohman
and Van Der Hamsvoort, 1998).

Conclusions

Market failure means that it is desirable for governments to intervene in the free
market-determined pattern of land use. Economic instruments work by setting
a price on the use of the environment, and rewarding agents for producing environmental goods. There are many advantages of this approach relative to the
alternatives, including flexibility in response, long-term adaptation and cost-
efficiency. We have shown how economic instruments might work in four
examples for Scotland.

However, it would be wrong to suggest that the wider use of economic instruments would be problem-free. Important questions include the extent to
which economic instruments are consistent with concerns over fairness, and
with the implied allocation of property rights. We already noted above that
spatial variability in vegetation recovery mitigated against a simple use of
tradeable stocking permits for red deer, if one is concerned with some target
level of environmental improvement, rather than simply a reduction in overall stocking rates at least cost. It seems likely that the best policy direction is
one which incorporates a *mix* of economic instruments, regulation, liability
and persuasion. None of the economic instruments put forward in our four
examples are designed to be stand-alone devices. For example, woodland
management will still require regulation in terms of ecological standards on
planting, even if at the same time higher payment rates are offered for higher
environmental-quality forests. Higher water quality for salmon in Scottish
rivers would be much assisted by taxing pollution sources or subsidising
bank-side re-vegetation, but other measures are required such as the prohibition of the use of dangerous substances in fish farming. Education and land
reform may be more important in reducing erosion in sensitive mountain
areas than increasing access costs.

It is also important to note the relatively modest objectives being stated for
wider use of economic instruments here: to improve environmental quality at a
lower cost to society than would be incurred absent such incentives. Neither are
we claiming that economic instruments (or, indeed, and other kind of instrument) can achieve the optimal pattern of land use. However, good policy
design involves making use of the full suite of available approaches, regulatory,
liability-based, voluntary and economic. Currently, the policy mixture is too
thin on the economic instruments side.

References

Baumol, W and Oates. W (1971). The use of standards and prices for the protection of the environment. *Swedish Journal of Economics*, **LXXIII**; 42–54.

Baumol, W and Oates, W (1975). *The Theory of Environmental Policy*. Cambridge: Cambridge University Press.

Bullock, C (1999). Environmental and strategic uncertainty in common property management: the case of Scottish red deer. *Journal of Environmental Planning and Management*, **42** (2), 235–252.

Gordon-Duff-Pennington, P (1997). The price may be deer. *Landowning in Scotland*, **245**; 23.

Hanley, N D (1989). Valuing rural recreation benefits: an empirical comparison of two approaches. *Journal of Agricultural Economics*, September; 361–374.

Hanley, N Hallett, S and Moffatt, I (1990). Why is more notice not taken of economists' prescriptions for the control of pollution? *Environment and Planning A*, **22**; 1421–1439.

Hanley, N D and Ruffell, R (1993). The Contingent Valuation of Forest Characteristics: Two Experiments. *Journal of Agricultural Economics*, May.

Hanley, N D and Spash, C (1993). *Cost-Benefit Analysis and the Environment*. Cheltenham : Edward Elgar Publishing.

Hanley, N and Sumner, C (1995). Bargaining over common property resources: applying the Coase Theorem to red deer in the Scottish Highlands. *Journal of Environmental Management*, **43**; 87–95.

Hanley, N, Kirkpatrick, H, Oglethorpe, D and Simpson, I (1998). Paying for public goods from agriculture: an application of the Provider Gets Principle to moorland conservation in Shetland. *Land Economics*, February; 102–113.

Hanley, N, Whitby, M and Simpson, I (1999). Assessing the success of agri-environmental policy in the UK. *Land Use Policy*, **16** (2); 67–80.

Hanley, N, Wright, R and Adamowicz, W (1998). Using choice experiments to value the environment: design issues, current experience and future prospects. *Environmental and Resource Economics*, **11** (3–4); 413–428.

Hanley, N, Alvarez-Farizo, B and Shaw, D (2000). Rationing an open-access resource: mountaineering in Scotland. Mimeo, IERM, University of Edinburgh.

Lacatz-Lohman, U and Van den Hamsvoort, P (1998). Auctions as a means of creating markets for public goods. *Journal of Agricultural Economics*, **49** (3); 334–345.

MacMillan, D C (1993). Commercial forests in Scotland: an economic appraisal of replanting. *Journal of Agricultural Economics*, **44** (1); 51–66.

Macmillan, D C, Harley, D and Morrison, R (1998). Cost-effectiveness of woodland ecosystem restoration. *Ecological* Economics, **27**(3); 313–324.

Macmillan, D C and Duff, E I (1998). The non-market benefits and costs of native woodland restoration. *Forestry*, **71**(3); 247–259.

MacMillan, D (2000). An economic case for land reform. *Land Use Policy*. **17**(2000); 49–57. OECD (1989). *Economic Instruments for Environmental Protection*. Paris: OECD.

Reid, C T (1995). Environmental regulation through economic instruments: the example of forestry. *Environmental Law and Management*, **7**; 59–63.

Shortle, J, Faichney, R, Hanley, N and Munro, A (1999). Least cost pollution allocations for probabilistic water quality targets to protect salmon on the Forth Estuary. Sorrell, S and Skea, J (eds). *Pollution for sale: emissions trading and joint implementation*; 211–230, Cheltenham: Edward Elgar.

Shortle, J, Abler, D and Horan, R. (eds) (2000) *Environmental Policies for Agricultural Pollution Control*. Oxon: CAB International

Sorrell, S and Skea, J (eds). *Pollution for sale: emissions trading and joint implementation*. Cheltenham: Edward Elgar.

Watt, J, Bartels, B and Barnes, R (1999). Declines of west highland salmon and sea trout. Lochailort: Lochaber and District Fisheries Trust.

Wightman, A (1996). *Scotland's mountains: an agenda for Sustainable Development*. Perth: Scottish Countryside and Wildlife Link.

WWF, (1995). *Wild Rivers: Phase 1 technical paper*. WWF Scotland, Aberfeldy.

Annex One

"Because of the added complexities and uncertainties involved with forestry projects, the Government remains convinced that the UK's priority should be emission reduction rather than carbon sequestration. But the Government does recognise that forestry projects can provide environmental benefits and it is therefore considering whether such schemes should also be permitted within a domestic trading scheme. However, rigorous monitoring and verification of forestry projects would be crucial if they are to provide a credible source of tradable credits. The Government believes that the credits from forestry schemes should only be redeemable against carbon targets if, for example, the projects were consistent with forestry definitions and methodologies as set out under the Kyoto Protocol; the companies offering the schemes could demonstrate how the offsets had been achieved and could assess how much carbon had been saved and over what timescale; and a long term management strategy was in place."

(DETR (1999) "Tackling climate change" available at http://www.environment.detr.gov.uk/climatechange/tackling/index.htm)

Economic and Environmental Regulatory Reform: the Case of the Scottish Water Industry*

John W Sawkins and Valerie A Dickie

Introduction

Since the establishment of a new Scottish Parliament in 1999 much has been written on the political, social and economic differences between Scotland and the rest of the United Kingdom (UK). Variations in political and economic institutional practice and procedure, once tolerated, are now celebrated, with the result that alternative solutions to generic problems are being adopted across the country.

For one industry – water – the political developments have merely underlined the fact that there has never been a single UK policy covering the delivery and regulation of this essential service. Indeed in the early 1990s, whilst the Labour Party's devolution policy was still being fine-tuned, the Conservative Government's review of the Scottish industry led to the implementation of an institutional settlement quite different from that pertaining in England and Wales.

The most recent regulatory innovation within the Scottish water industry – unparalleled in other parts of the UK – was the appointment of the first Water Industry Commissioner for Scotland on 1st November 1999. The Commissioner's primary responsibilities include the promotion of the interests of water authority customers, and the giving of advice to the Scottish Executive on the level of water charges over a period of years. At first glance this new regulatory settlement appears to have much in common with that operating in England and Wales. However the institutional arrangements for delivering water and sewerage services differ in several fundamental respects north and south of the border, and the development of economic and environmental regulation has reflected these differences.

This paper seeks to inform the current policy debate, by highlighting some of the main institutional and economic issues raised by the recent changes in regu-

* A modified version of this paper is forthcoming in the journal *Utilities Policy*.

latory arrangements for the Scottish water industry. The paper is divided as follows. Following the introduction, the subsequent two sections outline and analyse the development of the regulatory regime, highlighting major legislative changes and noting remaining structural flaws. The next section then outlines a new regulatory policy agenda, suggesting ways in which further developments might promote more effective regulation. The final section concludes.

Regulation: early developments

In Scotland, the regulation of water and sewerage services is currently a function of several bodies including the Scottish Executive, the Water Industry Commissioner and the Scottish Environment Protection Agency (SEPA). In terms of the historical development of regulation within the industry, however, this marks a low point in the number of separate institutions involved in the regulatory function.

Historically, water and sewerage services were delivered and regulated separately. This piecemeal approach, however, hampered the establishment of comprehensive systems of water supply regulation in both urban and rural areas. By the end of the second world war Scotland had 210 separate water authorities comprising of 170 town councils, 32 county councils, 6 water boards and 2 non-statutory water companies. Whilst the Water (Scotland) Acts of 1946 and 1949 went some way to harmonising the powers and duties of water authorities, it was not until the Water (Scotland) Act 1967 that the fragmentation of the industry was addressed with the consolidation and transfer of water authority functions to 13 large regional water boards. The 1967 Act also made provision for the establishment of the Central Scotland Water Development Board (CSWDB); a body charged with the development of new sources of water for the bulk supply of two or more of the constituent regional bodies.

For sewerage, public health concerns motivated the various legislative initiatives. Local acts supplemented the inadequate national statutes, whilst the number of separate sewerage operations – and hence regulatory institutions – eventually grew to exceed that of water. The first piece of comprehensive legislation was the Sewerage (Scotland) Act 1968 which embodied much local authority 'best practice'. On the eve of the 1975 Scottish local government reorganisation, however, responsibility for sewerage services was shared among 234 separate local authorities, a structure that had changed little from that established by the Local Government (Scotland) Act 1929.

Environmental regulation during this period had begun as one of the responsibilities of local authorities supplying water and sewerage services, but came to be administered locally by 21 river purification authorities (9 river purification boards and 12 local authorities)[1]. The overall economic regulatory framework within which the various types of authority operated was set by central Government – the Scottish Office – but the power to set charges was devolved to the various local authorities.

[1] They administered the Rivers (Prevention of Pollution) (Scotland) Acts 1951 and 1965.

Under the terms of the Local Government (Scotland) Act 1973 the nine regional and three islands councils established on 16th May 1975 became responsible for water and sewerage services within their areas. These twelve new local authorities – the upper tier of local government – were required by the legislation[2] to supply wholesome water to every part of their area where a supply of water was required for domestic purposes and could be provided at reasonable cost. A further obligation was to provide public sewers to drain their areas of domestic sewage, surface water and trade effluent. Sewage was to be treated at dedicated works or in another appropriate way. As with water the regional and islands authorities were not required to do anything that was not practicable at reasonable cost.

The role of the CSWDB as a developer of bulk supplies was preserved, and responsibility for the prevention of pollution of inland and defined coastal waters on mainland Scotland was transferred to seven River Purification Boards (RPBs)[3]. In theory these independent environmental regulatory bodies were to ensure environmental regulations were not breached. In practice their working relationship with regional councils was so close that environmental regulatory activity was circumscribed by the politics of their relationship[4]. In the case of the Scottish islands, these regulatory responsibilities rested with the councils anyway.

Overall responsibility for the economic regulation of the industry post 1975 remained with the Scottish Office. It continued to set the macroeconomic framework within which the bodies were to work, but delegated some regulatory powers to regional and islands councils that set charges at a level high enough to cover annual expenditure. The councils were responsible for their actions to the local electorate.

In general the benefits of the 1975 reforms were considerable. The two branches of the industry – water and sewerage – were consolidated and brought together for the first time, thereby allowing latent economies of scale and scope to be exploited. Harmonisation of technical and managerial functions across authorities, and regional planning, facilitated the dissemination of best practice, and promoted the allocation of resources to previously neglected areas within individual local authority areas. However serious structural flaws remained within this new institutional configuration.

First, as mentioned above, the effectiveness of environmental regulation was diminished by the very close working relationship between the water departments of the regional councils and their counterpart River Purification Boards. Pollution discharge standards were often relaxed, and the vigorous pursuit of offending authorities by RPBs rather rare.

---

[2] Statutory duties relating to the supply of wholesome water and the provision of public sewers were set out in two main pieces of legislation: the Water (Scotland) Act 1980 and the Sewerage (Scotland) Act 1968. Other related matters were dealt with in supplementary legislation

[3] The Islands Councils fulfil the role of the RPBs in their respective areas.

[4] As non-departmental government bodies RPB membership comprised appointees of District and Regional councils and the Secretary of State for Scotland. They did not, however, own or operate water infrastructure assets.

Second, responsibility for economic regulation was divided between central and local government allowing potential conflicts of interest to arise. In relation to capital expenditure for example, central government set strict limits on the funds available for capital works. As public sector organisations without access to private capital markets, the council water departments could only hope to raise additional money by cutting costs or raising charges. This latter course of action was invariably closed off by local councillors who were responsible for setting charges, but were unwilling for whatever reason to sanction above-inflation price rises. Consequently infrastructure investment ran at a level well below that required to meet new European quality standards on drinking and waste water.

These two structural flaws were, arguably, symptoms of the deeper problem of confused political regulatory accountability within the industry. In theory the presence of democratically elected local councillors on the water service boards of regional authorities guaranteed a strong accountability link between consumers of the service and suppliers. In practice comparatively little interest was shown by local politicians in water and sewerage services across Scotland as a whole and it was unclear in what sense they were accountable to the local electorate for these services in particular. With some honourable exceptions councillors saw water subcommittees of the councils as less attractive and lower profile outlets for their talents than say education or transport. Given their other duties and responsibilities very few were able to devote the time to acquire the technical expertise necessary to enable them to challenge the decisions of their permanent water services staff. Furthermore, their dependence on the local electorate for their political power and their antagonism towards central government could result in politically driven, short term, parochial decision making, to the detriment of the long term interests of the consumers of the service. This high degree of local political involvement may partly explain why employment levels in the Scottish water industry remained more or less steady for twenty years, whilst those in England and Wales fell steadily[5]. Throughout the period capital investment in the industry was also at the mercy of Central Government's spending decisions. This was another potential source of conflict between local and national politicians.

Regulatory reform: 1996 and beyond

Following privatisation of the ten English and Welsh Regional Water Authorities (RWAs) in 1989, the Conservative Government signalled its intention to restructure the Scottish industry. In common with England and Wales there was a pressing need to invest heavily in the infrastructure to bring Scottish water and sewerage services up to European standards. Years of underinvestment had resulted in a situation in which it was estimated for Scotland in 1992 that £5 billion was required to be spent over 15 years to ensure

---

[5] And most rapidly when local authority representation on English and Welsh Regional Water Authority Boards was removed in 1983. See Tables 1 and 2.

compliance with European Directives on Drinking Water (80/778/EEC) and
Urban Waste Water Treatment (91/271/EEC)[6]. There was a need to reform the
environmental regulatory arrangements to completely remove this function
from the bodies responsible for the delivery of the service. The Government
also wished to limit increases in the Public Sector Borrowing Requirement and
therefore sought to restructure the industry in such a way as to lever in private
sector finance to deliver this large investment programme. In contrast to the
English and Welsh situation, however, where the large RWAs had been organised and run on the integrated river basin management principle since their
creation, there was anecdotal evidence in Scotland to suggest that the structure
of the industry inherited from the 1975 reorganisation remained too fragmented. Opportunities for exploiting considerable financial, technical and
managerial economies of scale were apparently being lost.

In November 1992 the Scottish Office issued a consultation paper 'Investing
for Our Future: Water and Sewerage in Scotland'. In it the Government outlined eight options for the restructuring of the Scottish water and sewerage
industry as part of the wider process of local government reform. These
options ranged from no change to full privatisation. During the consultation
period, media and political interest in the issue was intense and almost univer-
sally hostile to privatisation. The Scottish Office's own 'Summary of
Responses to the Consultation Paper "Investing for our Future" ' concluded
that 92% of the 4,834 correspondents did not specify a preference for any particular option laid out in the consultation paper, but that 94% favoured
retention of the services within public control, and only 1% were in favour of
privatisation. Subsequent opinion polls and widely publicised regional postal
ballots conducted throughout Scotland confirmed these findings. Thus there
was little surprise amongst political commentators when the Government
announced its intention to create three new public water authorities in a White
Paper published in July 1993.

The statute reforming the industry passed into law as the Local Government
etc. (Scotland) Act 1994. Under its terms responsibility for water and sewerage
services was transferred from the nine regional and three islands councils to
three new Public Water Authorities (PWAs)[7] on 1st April 1996. The Central
Scotland Water Development Board was abolished, and a new economic regulatory body, the Scottish Water and Sewerage Customers Council (SWSCC),
was established to protect the interests of consumers. The Secretary of State for
Scotland became responsible for the appointment of all PWA and SWSCC
board members. This effectively removed responsibility for these services from
local government control for the first time, thereby also removing one source of
potential political conflict.

New economic regulatory arrangements were put in place whereby the
SWSCC was responsible, in the first instance, for scrutinising and approving
PWA Charges Schemes. In the event of deadlock between these parties the Sec-
retary of State would adjudicate. The transfer of delegated responsibility for

---

[6] The Scottish Office 1992.

[7] The North, East and West of Scotland Water Authorities.

pricing to boards comprised of central rather than local government appointees removed one of the main obstacles to price rises. These were cushioned, however, by central Government which granted Transitional Sewerage Relief on domestic bills. Nevertheless, although overall price rises were reduced by this scheme, and although PWAs operated within strict aggregate price caps set by the Scottish Office, consumers in several areas of the country experienced double digit price rises as charge harmonisation across the new PWA regions was brought in. Tables 3 and 4 record and illustrate the increases in net water and sewerage charges for domestic customers by PWA area.

Money for large scale capital investment remained under the regulatory control of the Scottish Office, with the Secretary of State for Scotland fixing External Financing Limits on an annual basis. The amount of money available through this route, however, was restricted and water authorities were encouraged to bridge any funding gap with money from the Private Finance Initiative (PFI) and increased charges. Thus although privatisation of the industry had not been implemented, the principle of private sector involvement had been established[8].

Parallel developments in environmental regulation were embodied in the 1995 Environment Act. This led to the abolition of the RPBs on 1st April 1996 when the new Scottish Environment Protection Agency (SEPA) became responsible for environmental regulation in general. However, unlike its English and Welsh counterpart – the Environment Agency – SEPA's activities were more narrowly defined. Nature conservation matters, for example, remained the responsibility of its sister agency Scottish Natural Heritage, and the Scottish Office retained responsibility for drinking water regulation. Nevertheless as a national environmental regulatory body established by statute with a wider range of powers and responsibilities than the old RPBs, SEPA had the potential to distance itself from local political pressure and operate more independently and aggresively than its predecessors. With its coverage of land and air as well as water it was also well placed to consider environmental issues in a holistic way, taking into account the interaction of impacts on the total environment. Furthermore, SEPA was to provide industry with a one stop shop for environmental regulation.

As far as environmental duties were concerned, the three water authorities were required to balance their obligations to supply water and drain sewage with conservation, environmental protection and public access duties[9]. This was to be done in conjunction with the Secretary of State for Scotland.

Once again the benefits of reorganisation quickly became apparent. As with the 1975 reorganisation the new authorities quickly went about integrating and improving the separate management and operating systems inherited from predecessor authorities. The creation of larger authorities allowed further economies of scale to be realised quickly. In oral evidence given to the House of Commons Scottish Affairs Committee[10] two years to the day after reorganisation the three PWA chief executives quantified this effect. For example, on its

8 Table 5 lists the water and sewerage PFI projects as at 20/7/98.
9 Section 65 Local Government etc (Scotland) Act 1994.
10 1st April 1998.

first day of operation the East of Scotland Water Authority had an operating budget which was £1.1 million less than for the aggregated former authorities. By the end of the first year a further £6 million of savings had been made, and a target of a 26% reduction in operating costs (in real terms) had been set for the year 2001. Similar cost savings had been made in the North and West of Scotland Authorities.

These savings went some way to releasing additional money for capital investment, that was further invigorated through continued engagement in Private Finance Initiative (PFI) schemes. Under previous local government arrangements infrastructure investment planning was fragmented and invariably hampered by the inadequate amounts of finance made available to them by central and local government. In England and Wales, the 1989 privatisation of the RWAs proved to be the start of the largest programme of capital investment in its history. Consequently the Conservative Government sought to promote the same end in Scotland, where the new PWAs were of a sufficient size to maintain teams of specialist staff able to deal with the legal, administrative and financial complexities of PFI projects. Once in Government the Labour Party found the attractions of PFI irresistible, and encouraged the PWAs to increase the amount of private sector capital employed in the industry. Table 5 lists the size and status of the industry's PFI projects to date. This stimulus lifted overall levels of capital investment in the Scottish Industry [Tables 6, 7 and 8] and provided a means by which private sector expertise was accessed by the PWAs.

One of the qualitative improvements in industry performance from the consumer's point of view has been the increased transparency of the water authorities' operations since 1996. Before reorganisation differing accounting and reporting systems made the comparative analysis of operations problematic. Although minutes of the regional and islands councils' water subcommittees and other ad hoc performance related pieces of information were notionally available to all members of the general public, access to this information was difficult to secure and the results hard to interpret. In the informational vacuum Scottish water industry 'myths' were perpetuated by local and national politicians unable or unwilling to call into question the performance of public sector operators for whom they were ultimately responsible.

One of the favourite 'environmental myths' was the assertion that Scottish consumers enjoyed the highest quality drinking water in the UK – or indeed in Europe. Clearly, if this had indeed been the case there would have been little need for the massive programme of mains renewal and reconstruction currently being undertaken by the successor bodies. In fact water quality in Scotland, although good, was no better than that enjoyed by consumers in England and Wales [Tables 9a, 9b][11]. Commenting on the report "Drinking Water Quality in Scotland 1996" the Scottish Environment Minister, Lord Sewel summed up the situation as follows,

> The report published today, while not showing an overall improvement in performance, does show an improvement in some important aspects. However, the

---

[11] See also Tables 10 and 11.

equivalent report for England and Wales, published by the Drinking Water Inspectorate in June 1997, showed an overall level of compliance with the regulations at taps in England and Wales of 99.6 per cent compared to 98.6 per cent in Scotland. 99.2 per cent of tap samples in England and Wales met the crucial microbiological standard, whereas only 97.7 per cent of samples in Scotland met this standard.

[Scottish Office News Release 1645/97, 3rd November 1997]

The foundation for improved transparency and information flows was laid by the Local Government etc. (Scotland) Act 1994. This legislation required PWAs to produce Codes of Practice (section 66), Schemes of Charges (sections 76 and 77) and audited accounts (sections 87 and 88). Reporting and information systems of the twelve predecessor bodies were consolidated by the PWAs and the task of public relations was professionalised in a way not seen before. The tangible effects of this included the publication of regular press releases, information leaflets and annual reports and accounts by each water authority. Public meetings of the water authority boards were advertised in the national press and systematic customer consultation became a feature of their operations. In addition PWA chief executives have appeared before the House of Commons Scottish Affairs Select Committee to respond to MP's questions on the public record.

This culture of increased openness, if anything, improved rather than diminished the authorities' public accountability, which now, following devolution, runs through the Scottish Executive to the Scottish Parliament (see below). Systems of official scrutiny, in addition to those outlined above, were opened up through the establishment of the SWSCC, its Area Committees[12], and the transfer of some – but significantly not all – of the economic regulatory functions from the Scottish Office. Freed of the need to adhere to a parochial political agenda the SWSCC was able to operate as a quasi-independent regulatory body, and used its limited powers to obtain some comparative information from the authorities. This mode of operation was not open to regional councillors – the local economic regulators – in the pre-1996 industry.

Another related benefit of reform was the commitment on the part of the water authorities to become increasingly customer orientated. Regular reporting and liaison with customers and their watchdog, the SWSCC, facilitated this. But it was clear that there was a shift in the mindset of management within the industry away from producer-driven priorities – the 'producer knows best' culture – towards customer-driven priorities. Evidence for this may be found in the East of Scotland Water Authority's customer strategy. A press release of 26/6/98 articulated this in the following way,

> The strategy will systematically change the values and culture of ESW to embrace not only customer care but also customer awareness and commitment. The key drivers behind the strategy are: customers define quality, not ESW, the provider; ESW will develop customer distinct competencies for all employees; the ultimate goal is customer loyalty, not acceptance.

[East of Scotland Water Press Release 26/6/98]

---

[12] One for each of the PWAs.

In an industry which remained dominated by engineering professionals with long experience of working in the public sector this was a radical and significant development. The momentum for this change was not all internally generated, however. Legislation, political initiatives such as the Scottish Affairs Committee scrutiny and the activities of the SWSCC in making comparisons with the PWAs' English and Welsh counterparts, all added to this development.

The establishment of a national environmental regulatory body also yielded real benefits in terms of the effective coordination and integration of environmental policy across Scotland. The enhanced status of the environmental regulator, SEPA, made it a more effective and influential body than the individual RPBs; and therefore able to influence policy formulation more effectively. It quickly sought to identify information requirements and survey the current state of the environment – an important calibration exercise. For example, in SEPA's first 'State of the Environment Report' (1996) the Chief Executive, Alasdair Paton claimed "building on the work of more than 60 predecessor bodies, the Agency is able to assess environmental quality as a whole and see the gaps in our knowledge of the impact of human activities." (SEPA 1996, p ii.) This claim was given substance in the context of water with the publication of "Improving Scotland's Water Environment" (SEPA 1999) which not only described the condition of Scotland's water environment, but also defined SEPA's targets for protecting and improving it.

Flaws in the settlement

Despite these gains, the settlement contained two serious structural flaws relating to the economic regulatory arrangements. First the SWSCC was given responsibility for considering charges just one year ahead. This perpetuated a short-termism that had hampered investment and operational planning within the industry for many years. The level of uncertainty pervading corporate plans and budgets was therefore much higher than that experienced by water companies south of the border who were given new price limits every five years.

A second more serious flaw lay in the division of economic regulatory duties between the SWSCC and the Scottish Office: the former being responsible for price regulation in the first instance and the latter retaining functions relating to the assessment of water company efficiency and capital expenditure requirements[13]. The Secretary of State, of course, remained the final arbiter of price rises.

The arrangement worked tolerably for the first couple of years during which agreement was reached between the PWAs and the SWSCC on price rises. In January 1998 the SWSCC, unable to reach agreement with the PWAs, referred the draft charges schemes to the Secretary of State. After consideration the draft charges schemes were accepted, with minor amendments, thereby calling into question the ability of the SWSCC to protect consumer interests. In effect the PWAs were able to 'second guess' their front line economic regulator by

---

[13] The Scottish Office remained responsible for setting External Financing Limits.

appealing to the Secretary of State for Scotland. This inevitably undermined the SWSCC's credibility and moral authority as a regulator. It also highlighted the need for a more powerful, politically independent economic regulator which combined the functions of the SWSCC with those of the Scottish Office.

Review and reform

Amongst the Labour Government's election promises was a commitment to conduct a thorough review of the Scottish water industry. On 16th December 1997 the Secretary of State for Scotland announced its outcome. Overall it proposed that the 1996 institutional structure should be left intact, that water and sewerage services should continue to be delivered by three public water authorities, but that a greater number of local councillors should be appointed to their boards. Crucially it recommended the integration and transfer of responsibility for various strands of economic regulation and price setting to a new Water Industry Commissioner.

Implementation of the review's recommendations was delayed in order to obtain ratification by the Executive of the new Scottish Parliament. This was forthcoming, meanwhile the necessary legislative changes were contained in the Water Industry Act 1999 which passed into law in June 1999. This statute established the post of Water Industry Commissioner for Scotland, dissolved the SWSCC and created three new Consultative Committees.

Under the terms of the statute the Water Industry Commissioner's primary function is to promote the interests of the water authorities' customers. In this he is to be advised by the three local Consultative Committees whose meetings he chairs. His specific duties include: advising the Scottish Executive on the level of water charges over periods of several years, investigating water customers' unresolved complaints, advising the Executive on standards of service and customer relations and approving the water authorities' codes of practice. In preparing his advice to the Executive he is required to have regard to the economy, efficiency and effectiveness of authorities in using their resources – a responsibility previously with the Scottish Office. However the responsibility for approving charges remains with the Secretary of State. If the Secretary of State rejects the Commissioner's advice he must publish his reasons for doing so. For the first time Scottish water authorities will be issued with individual price caps or limits on annual price increases. The Commissioner, like the SWSCC before, is accountable to the Scottish Executive and through it to the Scottish Parliament.

As far as environmental regulation is concerned the Commissioner is required to follow the lead given by Government ministers who frame environmental policy and determine priorities. Liaison with SEPA, the Scottish Executive and other bodies is an important part of the work. Nevertheless the Commissioner's regulatory remit is essentially an economic rather than an environmental one.

Given the reluctance of Government politicians to hand over ultimate responsibility for price setting to the Commissioner, the question of regulatory independence became pertinent. In general terms the Commissioner has been given 'operational independence'. In a Parliamentary statement Sarah Boyack, the Scottish Environment Minister, outlined the Executive's understanding of the term.

> We are committed to ensuring that the processs has maximum transparency. There will be no question of the commissioner's professional expertise being compromised or influenced by ministers. The legislation guarantees that all stages of the process, including both the commissioner's advice and ministers' decisions will be made public. That demonstrates our commitment to a process that is rigorous and open, and our commitment to avoiding short-termism.
>
> [Sarah Boyack, Scottish Parliamentary Statement 16 September 1999, col 566–7]

As in England and Wales where the Secretaries of State for the Environment and Wales determine overall investment needs and merely require the economic regulator to ensure that investment programmes are delivered efficiently, so in Scotland the Executive would determine investment in the light of legislative requirements.

> The commissioner's job will be to provide expert economic analysis and advice, but it is not his job to decide which areas of the water authorities' plans and operations are essential or optional. Those are issues for ministers, who are responsible for defining the standards of water quality and environmental protection that must be met by the water authorities. Most of those standards flow from European commitments, while others reflect the Government's own priorities for the industry. Therefore, the Scottish Executive will give the commissioner a statement defining the standards that must be met by the water authorities. The commissioner will still be able to challenge the cost of the work associated with those standards, but he will not be able to question the need for that work.
>
> [Sarah Boyack, Scottish Parliamentary Statement 16 September 1999, col 566]

The procedure is as follows. Government ministers first determine investment requirements in the light of their environmental policy commitments[14]. The Commissioner then makes an assessment of water authority financing requirements and submits his advice to Ministers. Should Ministers accept this advice it becomes binding on the water authorities. Should they choose to reject it they are required to publish their reasons for doing so. Finally, annual price caps are set, i.e. limits on the annual price increases that water authorities could levy during the period covered by the advice. The Commissioner's first advice covering the years 2000/1 and 2001/2 was released into the public domain in January 2000. Although broadly accepting the recommendations of his report, the Minister modified proposals for price limits in a downward direction.[15]

---

[14] These include: improving standards of urban waste water treatment by the end of 2000, bringing Scotland's designated bathing beaches up to European standards and investing to raise the quality of Scotland's drinking water.

[15] The Commissioner advised the Minister that charge increases for the next two years should be as follows: East of Scotland 19.9% and 14.9%; West of Scotland 19.9% and 14.9%; North of Scotland 35% and 27%. The final determinations were: East and West of Scotland 15% and 12%; North of Scotland 35% and 12%. News Release SE0171/ 2000 26 January 2000.

This early action by the Scottish Executive illustrates the degree to which they have retained powers relating to what may be considered to be microeconomic regulatory issues such as pricing and capital investment programmes. Although, under the United Kingdom's new constitutional settlement, water industry regulatory policy is a devolved power for the Scottish Parliament it would appear that the Executive has been reluctant to pass ultimate regulatory control on to the Water Industry Commissioner. As the new arrangements 'bed down' it will become clear to what extent the Commissioner's office will be able to influence the implementation of industrial policy for water, and to what extent officials within the Scottish Executive will continue to negotiate directly with water authorities over matters of pricing and investment, thereby bypassing the body set up to be the champion of the consumer.

Towards a new regulatory policy agenda

In general terms the 1999 regulatory settlement promises to improve upon previous arrangements by addressing the two main structural deficiencies of its predecessor outlined above: namely short termism and the division of economic regulatory responsibilities. Instead of annual price negotiations Ministers will request advice from the Commissioner over charges covering a period of years. With a more predictable income stream the planning function of water authorities will be streamlined. The Commissioner will also be responsible for the assessment of water company efficiency and will be in a position to challenge water authority estimates of operating and capital expenditure requirements and thereby recommend more challenging price limits to Ministers.

Regulatory operational independence

The legal requirement to publish price recommendations will also reinforce a trend, already apparent in the Scottish water industry, towards greater transparency. It remains unclear, however, whether the Commissioner will be able to exercise operational independence to the degree implied by early Government policy statements. All Government appointed regulators function within a political context of course; however political intervention is often more heavy handed in some industries than in others.

Historically local politicians have had much influence within the Scottish water industry. As noted earlier this may be one of the reasons why the industry remains overstaffed compared with its partners in England and Wales. And despite Government assurances of increased regulatory independence the potential for heavy handed political intervention remains. For it is Government politicians who appoint the Commissioner and his board members, Government politicians who reserve the right to make the final decisions over price rises and Government politicians, in the first instance, to whom the regulator is accountable.

Although the potential, the propensity and the means for political intervention exist within the industry, regulatory operational independence need not be an unattainable goal. As with the English and Welsh economic regulator, the Commissioner's office and duties are established in statute which gives him a measure of protection from overt political interference. The danger remains, of course, that further legislation or relentless Ministerial 'guidance' will compromise this position, therefore the Commissioner must be alert to the possibility of Ministers straying into areas which are properly his responsibility.

Therefore an early priority for the Scottish Commissioner should be to assert his independence and thereby establish and enhance his authority in the eyes of his regulatees and the public.[16] To this end one of the most powerful tools at his disposal will be his relationship with the press and media, and the way in which evidence on comparative PWA performance is fed into the public debate.

A useful early 'litmus test' of the maturity of the devolved Scottish political settlement may be the degree to which the Scottish Water Industry Commissioner's operational independence is guaranteed by Government Ministers.

Accountability and corporate governance

Operational independence, however, does not imply lack of political accountability. For clearly the Commissioner, as a public appointee, is not a free agent, and must therefore be properly accountable under the law for his actions. That accountability link currently runs through Ministers in the Scottish Executive to the Scottish Parliament.

The potential dilution of political accountability was one of the main arguments leveled by local government opponents to the 1996 reforms. However the removal of the industry from local government control has, on the whole, proved to be a liberating experience for managers in the industry, and it is to the credit of the Government that – in the face of fierce political opposition – its Review did not completely reverse the changes implemented in 1996. Nevertheless, it was recognised that greater operational independence should be accompanied by greater accountability, and for that reason the Government increased the numbers of locally elected councillors on each of the boards.

Doubts remain, however, over whether a higher proportion of local councillors sitting on water authority boards implies greater public accountability. Under the pre-1996 arrangements this accountability link was opaque as the main election platforms of local councillors seldom contained reference to local water issues. Other issues such as education or transport generally headed their political agendas and consumed most of their time and energies. Arguably, a PWA board, with its members selected carefully and made responsible to the Scottish Executive, and through it to Parliament promises equally effective democratic control.

---

[16] The first Water Industry Commissioner for Scotland is Mr Alan Sutherland. He took up post on 1st November 1999.

Ultimately arguments over the number of local councillors with seats on water authority boards are rather sterile, and there is a danger that appearance is put over substance when it comes to this issue. What matters is first, whether there is a fair cross section of civic society and opinion represented on these bodies, and second, whether there are systems in place to allow consumers to communicate with their representatives. The principle of accountability is a good one. But the new Scottish Executive and Parliament would be wrong to believe that they, or indeed local government politicians, have any monopoly right to the corporate governance of this industry. Arguably a 'leavening' of objective outsiders on every board is an essential element of a publicly accountable Scottish water industry. And, hand-in-hand with this must come a commitment on the part of board members to be aware of the preferences and opinions of their 'constituents', whether they be voters or the disenfranchised members of society.

Capital investment, environment and efficiency

Finally, the issue of capital investment will remain high up the industry's regulatory policy agenda for the foreseeable future. The need for a large programme of capital investment to improve the overall quality of the water environment in Scotland has already been discussed. The PWAs currently have plans for investing £1.7 billion over the next three years[17], and further projects will be required to raise the quality of Scotland's drinking water, sewage treatment systems and beaches to levels dictated by the various European Directives. In England and Wales the process of arresting and reversing many years of infrastructure under-investment and neglect began around the time of the privatisation of the RWAs in 1989. Table 8 illustrates the large increase in capital expenditure per head of population that occurred around that time. In Scotland, financing innovations such as Private Finance Initiative (PFI) projects are beginning to draw in the levels of capital required to turn the situation around.

As the industry's efficiency regulator, the Commissioner will have an important role in ensuring the capital investment programme is delivered at minimum cost. To date, anecdotal evidence suggests that cost estimates for operational and capital expenditure by the PWAs have not been challenged as vigorously as have those of the English and Welsh water companies by Ofwat. The division of regulatory responsibilities between the Scottish Office and the SWSCC meant that both bodies were at an informational disadvantage compared with their regulatory counterpart in England and Wales. Without robust regulatory challenges prices have risen at an unprecedented rate in Scotland, partly to accommodate increased capital expenditure. Thus since the 1996 reorganisation the average bill for Scottish domestic consumers, previously around a third of that charged to English and Welsh consumers, has more than doubled in nominal terms until it is now just over two thirds of the average

---

[17] Scottish Executive Press Release 1/7/99.

English and Welsh bill. And whilst English and Welsh customers are faced with bills falling in real terms in the year 2000, no end to the sharp Scottish price rises is in sight.

The position of vulnerable households in Scotland is of particular concern. The policy of regional charge harmonisation has undoubtedly exacerbated the problem in areas such as Forth Valley. However price rises of this scale for basic water and sewerage services have not been matched by changes in social security arrangements, which would undoubtedly mitigate the effects for disabled, elderly or low income households.

The main challenge for the Commissioner from the consumer's point of view therefore, is to bear down on prices as vigorously as possible whilst allowing PWAs to finance properly their large capital expenditure programmes. These programmes will generate the necessary improvements in environmental quality which have been missing from the Scottish industry for so long. An early priority for the new regulator will be to undertake research into potential efficiency improvements across industry activities. Useful comparative information such as standard costs for capital works as well as other research results are now available from the English and Welsh regulator. Consequently the Commissioner will be in a position to confront PWAs with this information and challenge them to justify any large variances which arise[18]. The use of a pseudo price-cap set for several years at a time, rather than annual rate of return targets will clearly improve the economic incentive structure facing the PWAs, and is likely therefore to promote overall efficiency.

Conclusion

The 1996 reform of the Scottish water industry set in motion a series of changes which have, already, benefited consumers of the service and are likely to lead to improvements in environmental quality. With the appointment of a new Water Commissioner for Scotland, consumers may now confidently expect vigorous economic regulation to accompany one of the largest programmes of environmental improvement via capital investment in the industry's history. The success of the Commissioner's activities will, however, hinge on whether the new Scottish Executive and Parliament give him adequate operational independence to regulate organisations which are amongst the largest natural monopolies still operating in the public sector.

Without this independence of regulatory action, consumers will lack an effective economic advocate willing to champion their cause. Government and industry will continue to set the industry's policy agenda, and consumers will remain rather passive recipients of whatever level of service these other stakeholder groups deem appropriate. At best a gradual slowing down of price rises will occur. However, long overdue debates, such as the question of designated support for low income households and the possible introduction of universal

---

[18] Comparative statistics suggest there is potential for cost cutting within the Scottish industry through reducing numbers of employees.

domestic metering in Scotland, may once again be stifled by a political and engineering establishment wishing to defend and preserve the charging status quo.

With powerful economic and environmental regulators, PWAs must find ways of meeting their environmental obligations without relying on double digit price rises to fund investment and improved customer service. They will instead have to discover ways of improving operating efficiency and environmental quality, whilst rising to the challenge of ever-higher public expectations. It remains to be seen whether the Scottish industry – publicly owned, managed and regulated – can deliver a level of service equal to that offered by its private sector counterpart in England and Wales.

Appendix.

Table 1   Number of Full-Time Equivalent Employees at March 31st.

Year	England and Wales	Scotland	Northern Ireland
1974/5	59,107*		
1975/6	60,649*		
1976/7	61,540*		
1977/8	61,837*		
1978/9	71,842		
1979/80	71,618		
1980/1	70,852		
1981/2	68,955	6,226	2,708
1982/3	66,540	6,121	2,605
1983/4	63,173	6,144	2,524
1984/5	59,606	6,129	2,518
1985/6	57,502	6,155	2,510
1986/7	56,774	6,182	2,605
1987/8	55,356	6,194	2,552
1988/9	54,575	6,094	2,569
1989/90***	53,908	6,162	2,584
1990/1	46,436**	6,229	2,598
1991/2	46,313**	6,514	2,633
1992/3	45,767**	6,608	2,533
1993/4	44,614**	6,665	2,463
1994/5	43,539**	6,653	2,418
1995/6	40,047**	7,004	2,308
1996/7	36,881**	6,890	2,260
1997/8	36,686**	6,463	2,206

Source: Waterfacts '84, Waterfacts '95, Waterfacts '98.

Notes:
*   Figures for England and Wales for 1974/5 to 1977/8 exclude employees of Statutory Water Companies.
**   Post 1989/90 figures exclude NRA staff. Most of these transferred from the Regional Water Authorities in 1989.
***   1989/90 figures are averages.
       1974/5 figure shown is number as at 31st March 1975 etc.

Table 2   Full-Time Equivalent Employees per Thousand Resident Population.

Year	England and Wales	Scotland	Northern Ireland
1974/5	1.19*		
1975/6	1.23*		
1976/7	1.24*		
1977/8	1.25*		
1978/9	1.45		
1979/80	1.45		
1980/1	1.43		
1981/2	1.39	1.20	1.76
1982/3	1.34	1.19	1.69
1983/4	1.27	1.19	1.63
1984/5	1.19	1.19	1.62
1985/6	1.15	1.20	1.61
1986/7	1.13	1.21	1.66
1987/8	1.10	1.21	1.62
1988/9	1.08	1.20	1.63
1989/90***	1.06	1.21	1.63
1990/1	0.91**	1.22	1.63
1991/2	0.90**	1.28	1.64
1992/3	0.89**	1.29	1.57
1993/4	0.87**	1.30	1.51
1994/5	0.84**	1.30	1.47
1995/6	0.77**	1.36	1.40
1996/7	0.71**	1.34	1.36
1997/8	0.70**	1.26	1.32

Source: Waterfacts '84, Waterfacts '95, Waterfacts '98. Annual Abstract of Statistics 1999, Table 5.1.

Notes:
*   Figures for England and Wales for 1974/5 to 1977/8 exclude employees of Statutory Water Companies.
**   Post 1989/90 figures exclude NRA staff. Most of these transferred from the Regional Water Authorities in 1989.
***   1989/90 figures are averages.

Table 3   Net Domestic Water and Sewerage Charges (£) (Band D).

| | 1996–97 | 1997–98 | 1998–99 | 1999–2000* |
	£	£	£	£
Water Authority Area				
North				
Tayside	81.50	113.76	158.52	204.60
Grampian	97.00	128.26	170.52	209.75
Highland	97.00	128.26	170.52	209.75
Western Isles	120.50	140.26	170.52	209.75
Orkney	120.50	140.26	170.52	209.75
Shetland	120.50	140.26	170.52	209.75
East				
Borders	95.50	122.79	157.23	186.00
Forth Valley	52.50	80.79	128.73	173.00
Fife	71.50	101.79	146.73	177.50
Edinburgh & Lothians	95.50	122.79	157.23	186.00
North Lanarkshire &				
East Dunbartonshire	92.00	118.43	155.33	186.00
Kinross	81.50	111.29	146.73	177.50
West				
Dumfries & Galloway	92.00	118.43	155.33	189.10
Strathclyde	92.00	118.43	155.33	189.10

Source: Scottish Water and Sewerage Customers Council, The Scottish Office.

Notes:
* Charges for 1999–2000 include gross charges for sewerage. The domestic sewerage relief grant ended in 1999.

Table 4 **Percentage (%) Increase in Net Domestic Water and Sewerage Charges (Band D).**

	Annual Change in Prices			Period 1996/97 – 1999/2000 Nominal Change in Prices %
	1996/97– 1997/98 %	1997/98– 1998/99 %	1998/99– 1999/2000 %	
Water Authority Area				
North				
Tayside	39.58	39.35	29.06	**151.00**
Grampian	32.23	32.95	23.00	**116.24**
Highland	32.23	32.95	23.00	**74.07**
Western Isles	16.40	21.57	23.00	**74.07**
Orkney	16.40	21.57	23.00	**74.07**
Shetland	16.40	21.57	23.00	**74.07**
East				
Borders	28.58	28.05	18.29	**94.76**
Forth Valley	53.89	59.34	34.38	**229.52**
Fife	42.36	44.15	20.97	**148.25**
Edinburgh & Lothians	28.58	28.05	18.29	**94.76**
North Lanarkshire & East Dunbartonshire	28.73	31.16	19.71	**102.17**
Kinross	36.55	31.84	20.97	**118.63**
West				
Dumfries & Galloway	28.73	31.16	21.74	**105.54**
Strathclyde	28.73	31.16	21.74	**105.54**

Table 5   The Scottish Water Industry PFI Project Progress as at 20/7/98.

Procuring Agency	Project Name	Capital Value £ million	Status	Financial Close (expected)
North of Scotland Water Authority	Inverness main drainage / Fort William sewage treatment	45	Signed	December 1996
North of Scotland Water Authority	Dundee, Carnoustie & Arbroath waste water treatment	100	Tenders invited / negotiation	October 1998
North of Scotland Water Authority	Aberdeen, Stonehaven, Fraserburgh & Peterhead sewage and sludge treatment	80	Potential	May 1999
Total		225		
East of Scotland Water Authority	Esk Valley purification scheme	21	Tenders invited / negotiation	December 1998
East of Scotland Water Authority	Almond Valley trunk sewer & Seafield sludge incineration	100	Tenders invited / negotiation	December 1998
East of Scotland Water Authority	Levenmouth purification scheme	47	Tenders invited / negotiation	Spring 1999
Total		168		
West of Scotland Water Authority	Dalmuir secondary sewage treatment	50	Tenders invited / negotiation	December 1998
West of Scotland Water Authority	Daldowie / Shieldhall sludge treatment centres	60	Tenders invited / negotiation	December 1998
West of Scotland Water Authority	Meadowhead, Ayr, Stevenston & Inverclyde sewage treatment	80	Potential	March 1999
Total		190		

Source: The Scottish Office.

Table 6   Total Capital Expenditure (£ million).

Year	England and Wales*	Scotland	Northern Ireland
1974/5	442.1		
1975/6	537.0		
1976/7	562.4		
1977/8	539.3		
1978/9	577.0		
1979/80	637.5		
1980/1	737.6		
1981/2	734.3		
1982/3	781.2		
1983/4	866.8		
1984/5	814.6	97.8	28.0
1985/6	908.0	95.9	25.2
1986/7	1046.1	104.6	25.8
1987/8	1196.0	99.5	25.3
1988/9	1338.9	112.4	25.2
1989/90	1896.7	127.4	40.1
1990/1	2570.7	144.5	44.7
1991/2	3195.1	171.1	51.4
1992/3	3099.7	230.4	64.6
1993/4	2855.3	262.0	82.0
1994/5	2524.9	259.9	97.1
1995/6	2625.2	238.2	102.9
1996/7	3385.8	225.7	76.3
1997/8	3663.5	282.3	–

Source:
England and Wales: Waterfacts '84 (for 1974/5–1978/9), Waterfacts '88 (for 1979/80 – 1983/4), Waterfacts '90 (for 1984/5 – 1988/9), Waterfacts '95 (for 1989/90 – 1993/4), Waterfacts '97 (for 1994/5 – 1996/7), Waterfacts '98 (for 1997/8)
Scotland and Northern Ireland: Waterfacts '90 (for 1984/5 – 1988/9), Waterfacts '95 (for 1989/90 – 1993/4), Waterfacts '97 (for 1994/5 – 1996/7), Waterfacts '98 (for 1997/8)

Notes:
* Figures sum expenditure by Regional Water Authorities (from 1989/90 the Water and Sewerage Companies) and Statutory Water Companies (from 1989/90 the Water only Companies). From 1989/90 some of the functions of the Regional Water Authorities were assumed by the National Rivers Authority (from 1996/7 the Environment Agency), these figures are also included in the totals. Two minor structural breaks occur in the series. From 1984/5 onwards expenditure on land drainage and related environmental projects were removed. From 1989/90 a change in accounting policy led to figures being prepared under infrastructure accounting assumptions.

Table 7   Capital Expenditure at 1997/8 Prices (£ million).

Year	England and Wales	Scotland	Northern Ireland
1974/5	2531.6		
1975/6	2475.1		
1976/7	2224.2		
1977/8	1841.0		
1978/9	1818.8		
1979/80	1772.2		
1980/1	1737.8		
1981/2	1546.5		
1982/3	1514.9		
1983/4	1607.1		
1984/5	1438.6	172.7	49.4
1985/6	1511.6	159.7	42.0
1986/7	1684.2	168.4	41.5
1987/8	1848.9	153.8	39.1
1988/9	1972.6	165.6	37.1
1989/90	2592.8	174.2	54.8
1990/1	3210.4	180.5	55.8
1991/2	3769.0	201.8	60.6
1992/3	3524.6	262.0	73.5
1993/4	3195.8	293.2	91.8
1994/5	2759.4	284.0	106.1
1995/6	2772.9	251.6	108.7
1996/7	3492.0	232.8	78.7
1997/8	3663.5	282.3	

Note:
Figures for 1997/8 are at cost terms. Actual expenditure in previous years has been repriced by RPI.

Table 8   Capital Expenditure at 1997/8 Prices (£) per Head of Population.

Year	England and Wales	Scotland	Northern Ireland
1974/5	51.2		
1975/6	50.0		
1976/7	45.0		
1977/8	37.2		
1978/9	36.8		
1979/80	35.8		
1980/1	35.0		
1981/2	31.2		
1982/3	30.5		
1983/4	32.3		
1984/5	28.9	33.6	31.9
1985/6	30.2	31.1	26.9
1986/7	33.6	32.9	26.5
1987/8	36.7	30.1	24.8
1988/9	39.1	32.5	23.5
1989/90	51.2	34.2	34.6
1990/1	63.1	35.4	35.1
1991/2	73.8	39.5	37.9
1992/3	68.7	51.3	45.4
1993/4	62.1	57.3	56.2
1994/5	53.5	55.3	64.6
1995/6	53.5	49.0	65.9
1996/7	67.1	45.4	47.3
1997/8	70.2	55.1	

Table 9a   England and Wales – Drinking Water Quality Compliance.
(Percentage of zones complying at all times with prescribed concentrations or values).

Parameters	England and Wales 1996	1997	1998
Total coliforms	98.8	98.5	99.4
Faecal coliforms	96.2	96.1	97.4
Colour	99.9	99.6	99.6
Turbidity	96.1	97.3	96.1
Hydrogen ion	98.3	98.4	98.2
Aluminium	97.2	97.7	98.3
Iron	76.1	76.9	79.0
Manganese	91.8	92.3	94.0
Lead	86.9	88.6	89.5
Total trihalomethanes	99.7	98.7	97.1

Source: Waterfacts '98, Table 3.10. Drinking Water Inspectorate, Drinking Water 1998, Table 8.4.

Table 9b   Scotland – Drinking Water Compliance 1997. Percentage of Determinations Not Exceeding Prescribed Concentration or Value (PCV) or Relaxed PCV.

Scotland – Water Authorities Parameters	North	East	West
Total coliforms	96.3	98.2	98.4
Faecal coliforms	98.1	99.7	99.7
Colour	94.3	99.9	98.1
Turbidity	99.4	99.7	99.4
Hydrogen ion	98.6	99.8	99.7
Aluminium	97.1	99.7	96.7
Iron	93.5	98.2	95.2
Manganese	97.5	99.3	99.3
Lead	97.8	99.9	98.5
Total trihalomethanes	74.3	75.3	60.0

Source: Drinking Water Quality in Scotland 1997, The Scottish Office.

Notes:
 The Secretary of State for Scotland authorised the following relaxations during 1997.
 North of Scotland (total 298 zones): relaxations for colour (198 zones), iron (159 zones), manganese (116 zones) natural aluminium (63 zones), oxidizability (128 zones), pH (37 zones), sodium (2 zones), chloride (2 zones), taste (11 zones), conductivity (4 zones) fluoride (1 zone) and odour (5 zones).
 East of Scotland (total 117 zones): relaxations for colour (19 zones), turbidity (12 zones) and iron (10 zones).
 West of Scotland (total 179 zones): relaxations for colour (62 zones), turbidity (11 zones), pH (4 zones), aluminium (17 zones), iron (53 zones), manganese (63 zones) taste (1 zone) and odour (1 zone).

Table 10   Sewage Treatment Works – Compliance with Discharge Consents (numeric).

	England and Wales		Scotland	
	number of works monitored	% monitored that comply	number of works monitored	% monitored that comply
1996	4,016	97	602	78.2

Source: Waterfacts '98, Table 4.8.

Notes:
 Data for Northern Ireland is unavailable.

Table 11   Bathing Water Quality Compliance (Directive 76/160/EEC) – Coliform Results. Percentage compliance.

	Number of Identified Bathing Waters								
	England and Wales			Scotland			Northern Ireland		
	pass	total	%	pass	total	%	pass	total	%
1989	304	401	75.8	16	23	69.6	16	16	100.0
1990	318	407	78.1	12	23	52.2	15	16	93.8
1991	312	414	75.4	15	23	65.2	16	16	100.0
1992	328	416	78.8	15	23	65.2	15	16	93.8
1993	332	419	79.2	18	23	78.2	15	16	93.8
1994	345	419	82.3	16	23	69.6	15	16	93.8
1995	379	425	89.2	19	23	82.6	15	16	93.8
1996	386	433	89.1	21	23	91.3	16	16	100.0
1997	397	448	88.6	18	23	78.3	14	16	87.5
1998	413	458	90.2	12	23	52.2			

Source: Waterfacts '98, Table 4.11. Environment Agency, Scottish Environment Protection Agency

Table 12   Average (Weighted) Domestic Water and Sewerage Charges (£).

	1995/96 £	1996/97 £	1997/98 £	1998/99 £	1999/2000 £
England and Wales*					
Anglian	275	282	284	288	272
Dwr Cymru	263	264	282	294	300
North West	195	202	221	234	246
Northumbrian	199	209	216	228	241
Severn Trent	189	194	208	222	228
South West	317	329	343	354	356
Southern	214	225	242	257	273
Thames	172	181	191	201	206
Wessex	234	243	253	266	268
Yorkshire	207	217	220	226	234
Average Charge	227	235	246	257	262
Scotland**					
North		93	123	156	195
East		77	104	140	171
West		83	107	138	170
Average Charge*		84	111	145	179

Source:
*   England and Wales: Ofwat Reports on Tariff Structure and Charges, Ofwat, Birmingham, 1995/6 – 1999/2000.
**   Scotland: 1996/7 – 1998/9 SWSCC, 1999/2000 PWAs.
***   Averages. Figures are derived by weighting the average council tax bill in each property band in each water authority area by the proportion of households in each band in each area.

Bibliography

Byatt, I C R (1999). *Checks, Balances and Competing Pressures – Looking Forward at the Role of the Regulator*, Director General's Address to the Centre for the Study of Regulated Industries, London School of Economics, Birmingham : Ofwat.

Environment Act (1995). Chapter 25, HMSO.

House of Commons Library (1998). Water Industry Bill, Research Paper **98**; 117, , London : House of Commons Library.

Local Government etc. (Scotland) Act (1994). Chapter **39**, London : HMSO.

Sawkins, J W (1994). The Scottish Water Industry: Recent Performance and Future Prospects, *Quarterly Economics Commentary*, Fraser of Allander Institute, vol **20**, no 1; 49–51.

Sawkins, J W (1998). The Restructuring and Reform of the Scottish Water Industry: A Job Half Finished. *Quarterly Economic Commentary*, vol **23**, no 4; 41–51, Fraser of Allander Institute.

Scottish Development Department (1980). *Water in Scotland: A Review*, HMSO.

Scottish Environment Protection Agency (1996). *State of the Environment Report*, Stirling : SEPA.

Scottish Environment Protection Agency (1999). *Improving Scotland's Water*, Stirling : SEPA.

Scottish Office (1992). *Water and Sewerage in Scotland: Investing for our Future*, The Scottish Office, Edinburgh.

Water Industry Act (1999). Chapter **9**: HMSO.

Water Industry Act (1999). Explanatory Notes to Water Industry Act : HMSO.

Water Authorities Association / Water Services Association / Water UK (various) Waterfacts.

Devolution and the Environment

Colin T Reid

Environmental law is a subject which has a long history, but has developed rapidly in recent decades.[1] Growing awareness of the damage being done to the environment has led to major initiatives in environmental regulation, addressing issues from those of global importance, such as climate change, to those of very local impact, such as litter. The legal mechanisms to give effect to such measures must operate at international, national, and local levels and provide means by which the steps agreed at wider levels can be adapted and implemented in order to operate on particular sites. As far as Scotland is concerned, this requires mechanisms which can fit within the existing legal and political structures and which can produce and implement policies addressing the peculiar stresses on Scotland's own environmental riches, whilst giving effect to initiatives determined at international, European Community and British levels. As in so many other areas, the realisation of devolution transforms the way in which the regulatory process will work.

The Scotland Act 1998 created both the Scottish Parliament and the Scottish Executive which together have wide powers to govern Scotland.[2] These powers are subject to limitations in many ways and an overriding power remains in the hands of the United Kingdom authorities. Nevertheless, the main responsibility for most environmental matters now falls within the competence of the new Scottish authorities, although the restrictions on their powers will also be significant in this field. The purpose of this paper is to explore how devolution might affect the development of environmental law in Scotland.[3]

---

[1] C. Reid, "Environmental Law: Sifting through the Rubbish" 1998 Juridical Review 236.

[2] See generally: A. Page, C. Reid & A. Ross, *A Guide to the Scotland Act 1998* (1999, Edinburgh, Butterworths); C. Himsworth & C. Munro, *The Scotland Act 1998* (1999, Edinburgh, W. Green); J. McFadden & M. Lazarowicz, *The Scottish Parliament – An Introduction* (1999, Edinburgh, T & T Clark).

[3] The devolution arrangements introduced for Wales by the Government of Wales Act 1998 are more limited, in that the Welsh Assembly has no powers to make primary legislation. Nevertheless, as noted below, so much environmental law is made in the form of delegated legislation (whether under very broad powers in an Act of Parliament or to give effect to EC measures) that there too devolution will have a major impact. The arrangements for Northern Ireland under the Northern Ireland Act 1998 also devolve wide environmental responsibilities.

Outline of the new arrangements

The first elections for the Scottish Parliament took place on 6th May 1999, with the Parliament and Executive assuming their powers on 1st July.[4] The electoral system involves an element of proportional representation, with electors voting both for a constituency member in the traditional way and separately for regional members, chosen from party lists on the basis of the regional vote, taking account of the constituency seats won.[5] This system fulfilled two of the predictions made for it before the election took place, enabling the return of candidates from minority parties, including one Green Party candidate,[6] and denying any one party an overall majority in the Parliament.

Under the Scotland Act 1998, the Scottish Parliament has the power to make laws on any matter within its competence (s.28). The power to make law includes the power to amend, repeal or replace existing Acts of Parliament from Westminster. However, it is expressly stated in the Scotland Act that the powers of the UK Parliament are not reduced by the creation of the Scottish Parliament, so that it remains possible for the UK Parliament to legislate for Scotland, even in areas within the competence of the Scottish Parliament (s.28(7)).

The competence of the Scottish Parliament is defined by providing that the Parliament has the power to deal with any matter apart from those expressly excluded from its competence.[7] There are some general limitations on the powers of the Parliament, and then a list of matters beyond its competence, which are designated in the Act as "reserved matters". The definition of reserved matters is complex. There are some general reservations and many other more specific ones, all with exceptions and some further exceptions to the exceptions; it takes 28 pages of the Act to define the reserved matters. Among the important general limitations (s.29), the Parliament cannot pass legislation which:

- will form part of the law of any territory outside Scotland;
- is incompatible with European Community law; or,
- is incompatible with rights under the European Convention on Human Rights.

The matters beyond the competence of the Parliament are set out in some detail. A number of enactments, most notably most of the Scotland Act itself and key parts of the European Communities Act 1972, are expressly protected from amendment or repeal by the Scottish Parliament (Sched.4). The reserved matters which remain in the hands of Westminster cover a wide range of issues,

[4] Scotland Act 1998 (Commencement) Order 1998, SI 1998 No. 3178.
[5] A. Page, C. Reid & A. Ross, *A Guide to the Scotland Act 1998* (1999, Edinburgh, Butterworths), chap.2.
[6] Robin Harper, MSP, who was returned for a regional seat in the Lothians.
[7] This is known as the "retaining model" of devolution, as opposed to the "transferring model" which operates by conferring on the devolved authorities only those powers specifically listed, as adopted in the abortive Scotland Act 1978 and in the Government of Wales Act 1998.

from immigration to the Ordnance Survey (Sched. 5). These reservations reflect the fundamental point that the new constitutional settlement is one of devolution within the United Kingdom, so that matters which are considered to be "more effectively and beneficially handled on a United Kingdom basis"[8] are excluded from the powers of the Scottish Parliament. These issues include the constitution, foreign affairs, defence, most economic and financial matters, social security, child support, abortion, equal opportunities, health and safety, company and competition law, intellectual property, broadcasting and many energy and transport matters.

Aside from political arguments over whether it is appropriate for certain matters to be reserved, inevitably there will be difficulties in determining the boundaries in some cases. Legislation of the Scottish Parliament which incidentally affects a reserved matter will be valid, but the Secretary of State (a member of the UK government) can intervene if he or she considers that such legislation will have an adverse effect on the operation of the law as it applies to reserved matters. The Secretary of State can likewise intervene if a Bill is incompatible with any international obligation or with the interests of defence or national security (s.35).

The creation of the Scottish Executive is as significant as that of the Scottish Parliament. The Executive is the government of Scotland, with responsibility for running the country on all matters within the competence of the Parliament. Their power therefore extends to health, education, local government, housing, home affairs, agriculture, fishing and forestry, sport, the arts, and many environmental and transport issues.

The Executive, collectively known as the Scottish Ministers, comprises the First Minister, appointed by the Queen on the nomination of the Parliament, and Ministers, appointed by the First Minister with the approval of the Queen which can be sought only with the prior approval of the Parliament (ss.44–49). Their powers extend to all ministerial functions (including the making of delegated legislation) in relation to Scotland which fall within the competence of the Parliament (ss.53–54). The Executive's powers actually extend further than those of the Parliament in that arrangements are made for "executive devolution", that is the transfer to Scottish Ministers of certain functions within the area of reserved matters (s.63). For example, although oil and gas pipe-lines and nuclear installations are reserved matters,[9] the power to grant consent for cross-country pipe-lines[10] and powers in relation to permits for the extraction of uranium and plutonium[11] have been transferred to the Scottish Ministers.[12]

Where powers are currently exercised on a United Kingdom or Great Britain basis, there is provision for the power to be disaggregated to allow the

---

[8] *Scotland's Parliament* (Cm.3658, 1997), p.10.

[9] Scotland Act 1998, Sched.5, Part II Sections D2 and D4.

[10] Pipe-lines Act 1962, s.1; a "cross-country pipe-line" is one over 16.093 km. long (ibid., s.66(1)).

[11] Nuclear Installations Act 1965, s.2.

[12] Scotland Act 1998 (Transfer of Functions to the Scottish Ministers etc.) Order 1999, SI 1999 No.1750, Sched.1.

devolution of its exercise in Scotland to a Scottish Minister (s.106 (1)-(2)). Certain powers are specified as ones to be exercised concurrently by UK and Scottish Ministers, and others can be added to the list (s.56). Public authorities which have both devolved and reserved functions can be designated as "cross-border public authorities", which means that they report to both Parliaments and that powers relating to their activities initially remain with the UK ministers, even in relation to devolved matters, but with requirements to consult the Scottish Ministers and the possibility of individual arrangements being made to meet the circumstances of each body (ss.88–90).[13] Examples of cross-border bodies include the Royal Commission on Environmental Pollution, the Joint Nature Conservation Committee, and the Forestry Commission.[14]

The overall effect of these provisions is to create a very complex web of statutory powers. In any statute where a power is granted to "the Secretary of State", it may take considerable effort to ascertain where the power now lies. If the subject matter is not a reserved matter, power lies with the Scottish Ministers. If it does fall within reserved matters, then it must be checked whether executive devolution has transferred at least part of the power to the Scottish Ministers, and there are cases where powers are exercised concurrently by UK and Scottish ministers and ones where the UK minister retains the power to act but only after consultation with Scottish Ministers. Electricity provides an example of this. The generation and supply of electricity is a reserved matter,[15] but the power to grant consents for generating stations and overhead lines in Scotland has been transferred to Scottish Ministers by means of executive devolution,[16] powers to require certain statistical information are exercisable concurrently,[17] while the major power to license suppliers of electricity remains with the Secretary of State but can be exercised only after consultation with Scottish Ministers.[18]

The Ministers are subject to the same limitations as the Parliament in that any action incompatible with European Community law or rights under the European Convention on Human Rights is invalid,[19] and there is the further limitation that the Secretary of State can intervene if any action is

---

[13] Scotland Act 1998 (Cross-Border Public Authorities) (Specification) Order 1999, SI 1999 No.1319; Scotland Act 1998 (Cross-Border Public Authorities) (Adaptation of Functions) Order 1999, SI 1999 No.1747.

[14] There have been many amendments to the Forestry Act 1967 to give effect to the transfer of responsibilities to the Scottish Ministers; Scotland Act 1998 (Cross-Border Public Authorities) (Adaptation of Functions) Order 1999, SI 1999 No.1747, Sched.12.

[15] Scotland Act 1998, Sched.5, Part II Section D1.

[16] Scotland Act (Transfer of Functions to the Scottish Ministers etc.) Order 1999, SI 1999 No.1750, art.2 and Sched.1.

[17] Ibid., art.3 and Sched.2.

[18] Ibid., art.4 and Sched.3.

[19] Litigation so far has centred on the actions of the Lord Advocate in prosecuting criminal cases in circumstances which are claimed to breach the accused's right to a fair trial under art.6 of the Convention, e.g. *HMA v Little* 1999 SLT 1145, *Starrs v Ruxton* 2000 SLT 42.

incompatible with any international obligation or with the interests of defence or security or might have an adverse effect on the law in relation to reserved matters (ss.57–58).

The limitations on the powers of the Parliament and the Executive are controlled in several ways. Where the courts are involved in any of the three jurisdictions in the United Kingdom, special procedural rules take effect in order to refer cases to appropriate courts, with the Judicial Committee of the Privy Council acting as the final court of appeal (Sched.6).[20] There are pre-legislative checks to ensure that proposed legislation by the Parliament is within its competence, including provision for the Law Officers[21] to refer the matter to the Judicial Committee (ss.31–33). The validity of any legislation or exercise of power can be challenged in the courts by means of judicial review, and the Law Officers[22] can refer to the Judicial Committee any "devolution issue" (essentially any question of whether legislation or governmental action falls within devolved competence (Sched.6 para.1)) even though it is not the subject of other legal proceedings. In addition to direct challenges, whether by private parties or as a result of a reference by one of the Law Officers, such issues can arise in any litigation, and the procedural rules provide for the reference of such issues to appropriate higher courts, and for the Law Officers to be notified. There would obviously be considerable disruption if a piece of legislation which has been relied on for some time is subsequently found to be invalid, and in order to limit this disruption, provision is made for the courts to remove or limit the retrospective effect of their decisions, or to suspend them to allow remedial action to be taken (s.102).[23]

The Secretary of State, a member of the UK government, also has wide powers to intervene. A Bill can be halted if he or she reasonably believes that it has provisions incompatible with the United Kingdom's international obligations or national security or has an adverse effect on reserved matters (s.35). Delegated legislation can be revoked on the same grounds and Scottish Ministers can be directed by the Secretary of State to take, or not to take, action to ensure that there is no incompatibility with international obligations (s.58). The UK Parliament retains full power to legislate for Scotland (s.28(7)). In order to provide a means of dealing with the practical problems which might arise when the

---

[20] Judicial Committee (Devolution Issues) Rules Order 1999, SI 1999 No.665.

[21] Here restricted to the Lord Advocate, the Attorney General and the Advocate General (the new Law Officer whose role is to provide legal advice to the UK government in relation to Scotland since the Lord Advocate who has fulfilled this role becomes a member of the Scottish Executive (s.87)).

[22] Here the Lord Advocate, the Advocate General, the Attorney General and the Attorney General for Northern Ireland.

[23] This option is not available where the challenge is to an "act" of the Executive, which is why the decision that it was in contravention of the European Convention on Human Rights for the Lord Advocate to prosecute cases before temporary sheriffs (as a result of their method of appointment and limited tenure in office) had to be given immediate effect, despite the major disruption to the criminal justice system; *Starrs v Ruxton* 2000 SLT 42.

legality of any law or action is in doubt, subordinate legislation can be made with such provisions as are necessary or expedient in the light of any legislation or action in Scotland which is actually or potentially invalid (s.107).[24] Used sensibly, this power enables the inevitable difficulties in any division of powers to be minimised, but does give the UK authorities a wide power to intervene if there is any question of the Parliament or Executive testing the limits of their powers.

Environmental Law in Scotland

The existence of the separate legal system in Scotland has meant that even before devolution there are differences between the law in Scotland and that in the rest of the United Kingdom. So much environmental law is of recent statutory origin, and derived from international or EC initiatives, that the distinctiveness of Scots environmental law is limited, but there can be issues of significant divergence. The most noticeable differences are perhaps those affecting the public authorities in this area. The Scottish Environment Protection Agency (SEPA) is not only quite separate from but has slightly different functions from the Environment Agency which operates in England and Wales.[25] In Scotland, SEPA is the regulatory body for all matters under Part I of the Environment Protection Act 1990, whereas in the south certain air pollution matters are in the hands of local authorities, whilst SEPA does not share the full flood protection responsibilities which its southern counterpart inherited from the National Rivers Authority. Scottish Natural Heritage fulfils functions which in England are divided between English Nature and the Countryside Agency.[26] The structure of local government is also very different in many respects; in Scotland (and Wales) there is a uniform system of single-tier authorities, whereas in England there is a mixture of unitary councils and two-tier county and district councils, with separate arrangements for London, to include a directly-elected Mayor. The administrative arrangements in Northern Ireland are quite different again.[27]

In substantive terms, the law may be found in separate places even when the content of the law is essentially the same; for example, although found in different statutes and associated regulations, planning legislation[28] in the two

---

[24] Such legislation can have retrospective effect (s.114).

[25] Environment Act 1995, Part I.

[26] Natural Heritage (Scotland) Act 1991, Part I.

[27] Although the end result is usually similar, on all issues the structure of the law in Northern Ireland tends to be very different from that in Great Britain; see generally S. Turner and K. Morrow, *Northern Ireland Environmental Law* (1997, Dublin, Gill & Macmillan). In the last few years notable efforts have been made towards the long-overdue implementation in Northern Ireland of EC Directives on environmental matters.

[28] Primarily the Town and Country Planning Act 1990 and the Town and Country Planning (Scotland) Act 1997.

jurisdictions in Great Britain is broadly parallel and case-law can largely be used interchangeably in both jurisdictions. Indeed, the recognition by the House of Lords in a Scottish case of "*Grampian* conditions" led to their use throughout Great Britain.[29] Even in modern statutory schemes there can be some differences reflecting not just stylistic preference among the legislators but areas where the law has responded to the varying context across the United Kingdom, e.g. the introduction in England and Wales, but not in Scotland of regulations controlling stubble burning[30] or the creation of the Advisory Committee on the designation of SSSIs introduced in relation to Scottish sites when Scottish Natural Heritage was created.[31] Some matters have no direct counterpart in the other jurisdictions, e.g. the absence of National Parks in Scotland.[32]

More significant, perhaps, are the differences in the broader legal background. Despite some convergence over the last few centuries, the separate historical development of the law in Scotland has led to many distinctive features when compared to the systems in England and Wales and in Northern Ireland which share a common history. The law of property, particularly in relation to land, is very different, with an obvious impact on all issues affecting interests in land. The law of nuisance imposes different tests for liability.[33] Procedural differences are also relevant, e.g. the procedures for seeking judicial

---

[29] *Grampian Regional Council v Secretary of State for Scotland* 1984 SC (HL) 58; the innovation here was that although it was recognised that the law did not allow conditions in the grant of planning permission to require the developers themselves to achieve results beyond their direct control (e.g. the provision of access over another's land), nevertheless a condition was declared valid which prohibited the commencement of development unless such results were achieved.

[30] The power to make regulations extends to Scotland but has only been exercised south of the border where the issue was seen as having higher priority because of the greater expanse of arable land and several incidents when drifting smoke had on created dangerous conditions on motorways; Crop Residues (Burning) Regulations 1993, SI 1993 No.1366, made under the Environmental Protection Act 1990 s.152.

[31] Natural Heritage (Scotland) Act, s.12; the background to this provision is that members of the House of Lords (including land-owners) seized the opportunity presented by the Bill to voice their dissatisfaction at the operation of the SSSI system, especially when applied to large areas of land and the eventual form of s.12 is a much watered-down from the measures first inserted into the Bill in the Lords (see annotation in *Current Law Statutes 1991*, 28/12 (pp.28–15 to 28–17)).

[32] When the legislation establishing National Parks was framed shortly after the Second World War (National Parks and Access to the Countryside Act 1949), it was felt that in the absence of such intense pressures from developers and visitors, there was not the same need in Scotland for the special arrangements devised for the Parks. It has recently been proposed that National Parks be created in Scotland (See note 74 below) and the National Parks (Scotland) Bill was introduced to the Parliament at the end of March 2000..

[33] In Scotland there is no concept of public nuisance, there must always be fault, *culpa*, and there is no equivalent of the strict liability under *Rylands v Fletcher* (1868) LR 3 HLC 330; see generally G. Cameron, "Civil Liability for Environmental Harm" in C.

review differ in relation to issues such as standing[34] and time-limits.[35] In particular, the criminal justice systems differ in many ways. Not only do the rules of evidence and procedure place very different demands on those prosecuting offenders, but crucially in Scotland all prosecution (except in most exceptional circumstances unlikely to arise in the environmental context) lies in the hands of the public prosecutor. Hence, whereas the Environment Agency in England and Wales can commence prosecutions itself, the Scottish Environment Protection Agency must refer cases to the procurator fiscal who then has sole control of whether and how to proceed.

Constraints on Divergence

The creation of the Scottish Parliament and Executive would suggest that in future the law in Scotland will develop in its own way, separately from that in the rest of the United Kingdom. There are, however, several major constraints which will limit the extent of any divergence in environmental law. The range of reserved matters which remain in the hands of the London authorities limits the scope for the Scottish Parliament to make major changes in many fields. The need to ensure compliance with EC law will also be a great restriction on the new authorities' freedom of action, whilst the similar requirement in relation to international obligations and human rights will also limit the scope for new initiatives. Each of these will be examined briefly.

The Scotland Act 1998 operates on the basis of the transfer to Scotland of all powers not expressly reserved to the UK Parliament and government. Various government publications have, however, provided summary lists of matters which will be within the competence of the Scottish authorities and the environment correctly features on such lists.[36] Yet a number of important areas are reserved and although the Scottish Parliament will gain control over many issues, e.g. town and country planning, water pollution, and integrated pollution control, some of the reserved matters are either significant in their own right or limit the extent to which any wholly integrated approach to environmental issues can be adopted by the new bodies.

The reservation of taxes and excise duties (with limited exceptions) will prevent the use of many economic instruments in the pursuit of environmental objectives (A1).[37] This is a particularly important point given the increasing prominence of fiscal measures in environmental policies. The landfill tax plays a central role in the government's strategies to limit waste production, with the charge on every tonne of waste disposed of in landfill sites being de-

---

[34] C. Reid, "Judicial Review and the Environment" in B. Hadfield (ed.), *Judicial Review: A Thematic Approach* (1995, Dublin, Gill & Macmillan) pp.41–48.

[35] C. Reid, "Forestry and Environmental Assessment: The Benefit of Hindsight" (1999) 11 JEL 177.

[36] E.g., *Scotland's Parliament* (Cm.3658, 1997), p.6.

[37] This reference and the ones following are to the Sections within Part II of Schedule 5 to the Scotland Act 1998, which set out the specific reservations.

signed to encourage all waste producers to reconsider both alternative disposal routes and ways of avoiding the waste being created.[38] Yet the Scottish authorities have no power over the level of the tax, nor even the detailed rules on liability, exceptions or the scope of payments for environmental purposes which will generate an entitlement to tax credits. More controversial was the UK government's policy of a steady increase in the road fuel duty as a key element of the strategy to meet international commitments to reduce the production of climate changing gases.[39] This policy, now abandoned, led to protests in the rural parts of Scotland, where there is no alternative to road transport and commercial factors already mean that fuel prices are higher than in urban areas. This, though, was a matter wholly outwith the power of the Scottish authorities.[40]

Other instruments for influencing behaviour are also not available to the devolved authorities. For example, initiatives designed to utilise consumer pressure, e.g. regulating environmental claims in advertising or imposing additional labelling requirements, are outwith the powers of the Scottish authorities in view of the reservation of matters affecting consumer protection (C7) and product standards (C8). Such reservations significantly reduce the range of mechanisms available to the Scottish Parliament to implement environmental policy.

The position in relation to energy and transport is particularly complex, and the division of responsibilities between Holyrood and Westminster will not ease the development of integrated policies seeking sustainable solutions to the difficulties faced in these areas. In all of the cases noted below, the pattern is for a large area to be reserved for Westminster, but with a number of exceptions and several matters within the reservation becoming the responsibility of the Scottish Executive through executive devolution. Electricity (D1), oil and gas (D2), coal (D3), nuclear energy (D4), and energy conservation (D5) are all reserved matters, but with exceptions covering a number of the pollution control tasks of SEPA[41] while further powers in relation to individual projects have been transferred to Scottish Ministers.[42] The difficulties faced in trying to implement sustainable policies are illustrated by the fact that although energy conservation is a reserved matter (D5), the Scottish authorities will be allowed to promote energy efficiency, but only through advice, publicity, grants, loans and other such positive devices, not by means of any taxes, prohibition or

---

[38] Finance Act 1996, s.53; see C. Reid, *Environmental Law in Scotland* (1997, Edinburgh, W. Green) pp.89–90.

[39] Kyoto Protocol to the United Nations Framework Convention on Climate Change (1997), (1998) 37 ILM 22.

[40] The matter was raised in the Transport and Environment Committee of the Parliament (Second Meeting of 1999, 8 September, 1999, cols. 35–36).

[41] In all cases the subject-matter of Part I of the Environmental Protection Act 1990 is excluded from the reservation and thus is within the competence of the Scottish Parliament and Executive.

[42] Scotland Act 1998 (Transfer of Functions to Scottish Ministers etc.) Order 1999, SI 1999 No.1750, art.2 and Sched.1.

regulation.[43] Similarly, in relation to transport, some road transport matters (E1), rail transport (E2), marine transport (E3), air transport (E4), and some other transport matters (E5)[44] are all reserved. Again, certain matters are excepted or are the subject of executive devolution. The development and operation of a wholly integrated transport policy will thus require great co-operation between the Holyrood and Westminster authorities.

The need to comply with European Community law imposes a major constraint on the Scottish authorities. Neither the Parliament nor the Executive is empowered to take action incompatible with Community law. The initial responsibility to ensure that EC law is implemented in Scotland will rest with the Scottish authorities – the very first Scottish Statutory Instrument made by the Scottish Executive was to implement an EC Directive[45] – but ultimately that responsibility rests with the UK government, and if a state is brought before the European Court of Justice to answer a charge of failure to comply with EC law, it is no defence to argue that the matter falls within the jurisdiction of some lower tier of government.[46] Accordingly, the UK authorities retain power to intervene if the Scottish authorities are not doing what is required by EC law, whether by taking measures contrary to EC requirements or by failing to give effect to EC provisions. In addition to the general illegality of action incompatible with EC law, the UK ministers retain the power to legislate for Scotland in order to implement EC obligations, under s.2(2) of the European Communities Act 1972 (s.57(1)). Where EC, or international, law creates an obligation to achieve certain targets, e.g. recycling of waste or reductions in greenhouse gas emissions, there is an express power to allocate a proportion of such targets to Scotland (s.106(5)).

The extent to which the need to comply with EC law constrains the Parliament may be felt all the more because there is no formal mechanism for the Scottish authorities to have a say in the formation of EC policy and law. It is the United Kingdom which is the member state of the Community and the UK

---

[43] This reservation also illustrates the complexity of working with the reservations. The actual reservation is of "The subject-matter of the Energy Act 1976, other than section 9", yet the content of section 9 (the use and liquidation of offshore gas) is actually covered by another reservation (D2, covering oil and gas). The exception for energy efficiency matters permits the Scottish authorities to undertake the "encouragement of energy efficiency other than by prohibition or regulation", yet certain additional mechanisms, such as taxes, controls on advertising or product labelling are beyond their powers because of other reservations (A1, C7 and C8 respectively).

[44] Including the transport of radioactive material and the carriage of dangerous goods.

[45] The Environmental Impact Assessment (Scotland) Regulations 1999, SSI 1999 No.1. The new authorities are proving as slow as their predecessors in implementing some EC measures. The EC Directive on integrated pollution prevention and control (Dir. 96/61/EC) was due to be implemented by October 1999 but as yet none of the necessary regulations, to be made under the Pollution Prevention and Control Act 1999, have been introduced.

[46] *Commission v Belgium* (Cases 227–230/85) [1988] ECR 1; *Commission v Italy* (Case C-33/90) [1991] ECR I–5987.

authorities which are entitled to a seat in the Council when decisions are being made. The way in which European Community matters are to be handled is therefore one of the important elements in the Concordats between the UK and devolved administrations (see below). Ultimately, though, the Scottish authorities are legally required to implement measures adopted at Community level without any legally guaranteed opportunity to have a say in their making. There may be some limited scope for independent action by the Scottish Parliament or Executive through the challenge in the European Court of Justice to the legality of Community legislation or decisions, but uncertainties remain over the status of regional governments in the Community legal order and the acceptability, legally and politically, of Scottish authorities acting independently of the UK government.[47]

Similar points can be made in relation to the need to avoid action incompatible with international obligations. In keeping with the general constitutional position that international relations is a matter for politicians, not judges, in this case the enforcement of the limitation lies in the hands of the Secretary of State, rather than the courts, but again this may limit the way in which the Scottish authorities can act. In particular, a strategy to comply with international obligations on such issues as climate change which might make sense in relation to the United Kingdom as a whole, may not be as appropriate for Scotland if it were viewed by itself, yet compliance in Scotland will be essential if the UK strategy is to succeed.[48] There is plenty of scope for disagreement between the two authorities.

A further limitation on the freedom of action of the Scottish authorities is the need to ensure compatibility with the European Convention on Human Rights. Although it will be October 2000 before the Human Rights Act 1998 is brought into force to give effect to Convention rights in other contexts, both the Scottish Parliament and the Scottish Executive must respect these rights from their inception. Neither has the power to act in a way which is incompatible with Convention rights, and legal challenges on this ground may be expected. In the environmental field potential areas of dispute are the extent to which environmental regulation fails to protect, or unduly interferes with, the rights to family life, home and property, and the fairness of decision-making procedures in regulatory systems.[49]

---

[47] N. Burrows, "Devolution and the European Union – Nemo me impune lacessit" – paper at the conference Legal Aspects of Devolution, Faculty of Laws and Constitution Unit, University College London, 23rd September 1999.

[48] The allocation of a target within the United Kingdom can be done only after consultation with the Scottish Ministers; Scotland Act 1998, s.106.

[49] See, for example, the arguments raised in *Powell and Rayner v United Kingdom* (1990) 12 EHRR 355 (whether aircraft noise amounted to an infringement of the right to respect for private life and home (art.8)), *Bryan v United Kingdom* (1996) 21 EHRR 342 (whether the appeal procedure in the town and country planning legislation provides an "independent and impartial tribunal" (art.6)) and *Buckley v United Kingdom* (1997) 23 EHRR 101 (whether refusal of a gypsy's application for planning permission to site a caravan amounted to a breach of respect for her home (art.8) and to

Initial Progress

Writing in March 2000, it is too early to reach any firm views of how the new
authorities in Scotland are addressing environmental matters, but several
points of note have emerged, relating to both structures and policies. As far as
structure is concerned, the lead has been taken by the Executive where a Minister has been appointed with responsibility for Transport and the Environment
– Sarah Boyack, who was a lecturer in planning.[50] Interestingly, these responsibilities straddle an administrative boundary within the Scottish
Administration.[51] Transport and planning matters fall within the hands of the
Scottish Executive Development Department, whereas environmental matters, including pollution control, are within the hands of the Scottish Executive
Rural Affairs Department (no matter how urban or industrial the issue). This is
a break from the standard Westminster pattern where ministerial portfolios
and departmental boundaries coincide, and offers the opportunity for greater
co-ordination and integration of policy provided that the inevitable difficulties
of working across departmental boundaries can be resolved.

The division of responsibilities within the Executive has been reflected in the
committee structure of the Parliament. The subject committees are intended
both to fulfil the scrutiny role carried out by departmental select committees at
Westminster and to play a major part in the preparation and consideration of
proposals for legislation. A Transport and Environment Committee has been
created, and it has heard evidence from the Minister on several occasions, covering the full range of her responsibilities. The Committee has referred to its
role as being not just to scrutinise what the Executive is doing but to draw its
attention to areas where it is failing to act. A wide range of issues has already
been raised before the Committee, including: National Parks, the national
waste strategy, smoke control areas, and environmental impact assessments.[52]
The report of its first detailed inquiry, on the extension of planning controls for
telecommunications developments, was published in late March 2000;[53] at that
time a second detailed inquiry, into concessionary fares for travel, was well
advanced. Other subject committees will also have a role in environmental

---

[49] (continued)   discrimination (art.14)); cases from other states are increasingly developing the law in environmental contexts, e.g. *Lopez Ostra v Spain* (1995) 20 EHRR 277
(severe pollution can be a breach of the right to respect for private and family life and
home (art.8)), *Matos e Silva Lda. v Portugal* (1997) 24 EHRR 573 (whether designation
of a nature reserve was a breach of the landowner's peaceful enjoyment of his posses-
sions (First Protocol, art.1)).
[50] The appointment of such a minister was one of the express commitments in the
coalition agreement between Labour and the Liberal Democrats (see below).
[51] The same applies to the Minister for Communities, whose responsibilities straddle
the Development and Justice Departments.
[52] The Committee met twelve times between June 1999 and March 2000.
[53] Transport and Environment Committee, Third Report 2000: *Report on inquiry into
the proposals to introduce new planning procedures for telecommunications
developments*.

issues, e.g. it is the Rural Affairs Committee which is taking the lead in considering the National Parks (Scotland) Bill.

In addition to the subject committees, there are other parliamentary committees which may prove influential. Given the extent to which environmental law is created through delegated legislation, the work of the Committee on Subordinate Legislation is significant.[54] In its early meetings it has already emphasised that its role relates not to policy but to procedural and drafting matters,[55] but several points which may have an influence on how the Executive legislates were already raised at its first few meetings. These include asking for more readily understandable notes on what the legislation is seeking to achieve, requesting fuller descriptions of any consultation procedures and seeking explanations of why legislation to implement European Community Directives is being made after the due date for implementation, why it has been thought necessary for implementation dates in Scotland and England to coincide and why legislation has not been made available to the Parliament sufficiently far in advance of the date when it comes into force.[56] The Transport and Environment Committee has also subjected delegated legislation within its remit to detailed examination, e.g. in its consideration of new regulations on environmental assessment it heard from witnesses from the Scottish Executive Development Department and the Forestry Commission.[57] The making of environmental law will be transformed if there is this kind of thorough parliamentary scrutiny of subordinate legislation.

The arrangements for co-ordination and consultation between the authorities in London and Edinburgh have also been addressed. The "Concordats", published in October 1999,[58] set out the relationship between the UK government and the devolved governments in Scotland and Wales, but it is expressly stated that these documents do not create binding legal obligations, but are "intended to be binding in honour only".[59] The Concordats reflect the

---

[54] This Committee is created and its remit set out in the Standing Orders of the Parliament; Scotland Act 1998 (Transitory and Transitional Provisions) (Standing Orders and Parliamentary Publications) Order 1999, SI 1999 No.1095, Rule 10.3.1.

[55] See, for example the Convener's opening words at the second meeting of 1999 of the Committee (31 August, 1999) col.9.

[56] See, for example, the report of the fifth meeting of 1999 of the Committee (21 September, 1999) cols.27–36.

[57] Third meeting of 1999 (22 September, 1999) cols.51–65, considering the Environmental Impact Assessment (Scotland) Regulations 1999, SSI 1999 No.1, and the Environmental Impact Assessment (Forestry) (Scotland) Regulations 1999, SSI 1999 No.43.

[58] *Memorandum of Understanding and supplementary agreements between the United Kingdom Government, Scottish Ministers and the Cabinet of the National Assembly for Wales*, Cm.4444 (1999); the main Memorandum is supported by an Agreement on the Joint Ministerial Committee and Concordats on Co-ordination of European Union Policy Issues, Financial Assistance to Industry, International Relations and Statistics, as well as by agreements with specific departments of the UK administration.

[59] Ibid., pp.3, 18, 28, 39.

fundamental nature of the new constitutional arrangements as devolution within the United Kingdom, with the UK Parliament legally entitled to address devolved matters and the UK government alone having responsibility for international relations and relations with the European Union. At the core of the Concordats lies a commitment to good communication, ensuring that the administrations have the opportunity to make representations to each other in sufficient time for these to be fully considered.[60]

A Joint Ministerial Committee, chaired by the Prime Minister, will meet at least once a year to discuss the arrangements for liaison between the administrations and resolve any disputes which cannot be resolved on a bilateral basis. The Committee may also meet in "functional format" (e.g. the agriculture or environment ministers) as required, always chaired by the UK minister, and in particular the Committee, chaired for this purpose by the Foreign Secretary, will be the principal mechanism for consultation on the UK position on European Union matters, although timescales may require this work to be done largely by correspondence. A committee of officials from the various administrations will operate to prepare for the meetings of the main Committee. The Committee is described as "a consultative body not an executive body, and so will reach agreements, rather than decisions",[61] but although the administrations are expected to support the agreed positions, they remain free to determine their own policies.

On European Union matters,[62] the emphasis is on the early provision of information to the devolved administrations to allow their full involvement in the formulation of the UK position. A single UK negotiating line will be developed and ministers and officials from the devolved administrations will work as part of the UK team, with the UK minister retaining overall responsibility. The UK minister will decide on a case-by-case basis which ministers will attend Council meetings and it will be possible for ministers from the devolved administrations to speak for the United Kingdom, putting forward the agreed policy position. There is thus no scope for the Scottish Executive formally to argue a separate case before the European institutions, although the devolved administrations may take part in informal discussions with European bodies, and indeed the Scottish Executive has established an office in Brussels for this purpose.

The implementation of EC measures within devolved competence is to be discussed by the UK and devolved departments, with the options of separate implementation or legislation for the whole of Great Britain or the United Kingdom. Again, the UK departments will take the lead in co-ordinating action. If action is taken against the United Kingdom for alleged breaches of EC law in a matter that is wholly within devolved responsibilities, the draft reply will be prepared by the devolved administration before being agreed with the UK department; otherwise the UK department will take the lead in

---

[60] Ibid., p.3.

[61] Ibid., p.10.

[62] Ibid., pp.18–23; the provisions on international relations are broadly similar, again giving effect to the UK government's legal responsibility for this area (pp.30–38).

consultation with the others. If any financial cost or penalty is imposed on the UK for a breach of EC law within the responsibility of a devolved administration, that administration will be responsible for meeting that cost. Any disputes are to be resolved bilaterally if possible, but then referred to the Joint Ministerial Council.

The Concordats provide a commitment to, and basic framework for, co-operation between the UK authorities and those in Scotland and Wales, but inevitably cannot guarantee satisfaction on all sides. Much will depend initially on the UK departments' appreciation of the extent to which responsibilities are now shared and of the ways in which their policies and decisions can have an impact on devolved matters – issues where the devolution issues are overlooked are perhaps likely to cause more frustration than those where there is an acknowledged policy difference. The extent to which the Scottish authorities can be reassured that their concerns are properly taken into account, especially at European and international levels, will also be significant. Ultimately, though, the political, and administrative, will to make the new arrangements work will be more important than any structural mechanisms.[63]

On policy matters, the environment has had a high profile in the early days of the Parliament, although it remains to be seen what becomes of this in practice. "Environment and Transport"[64] appears as one of nine headings in the coalition agreement between the Labour and Liberal Democrat parties which led to the establishment of the current Executive.[65] The appearance of such a heading is perhaps influenced by the fact that it was probably fairly easy to reach agreement in this area,[66] especially since most of the points were either existing policy (e.g. the establishment of National Parks in Scotland), not controversial (e.g. the appointment of a Minister for Environment and Transport) or made no specific commitment (e.g. the promotion of the use of renewable energy, or the aim for steadily improving standards of water or air quality). Nevertheless, the Agreement does provide a standard against which the achievements of the Executive can be judged, although measuring success may be difficult in some cases,[67] and some of the points are challenging. In particular, the opening commitment to "integrate the principles of environmentally and socially sustainable development into all government policies" and the promise to introduce Strategic Environmental Assessment for government programmes do call for significant changes in thinking across the entire government.

---

[63] One sign of the recognition of the new constitutional framework is the very title of *The Air Quality Strategy for England, Scotland, Wales and Northern Ireland* (Cm.4548, SE 2000/3, NIA 7).

[64] It is not clear whether the change from the promise of a "Minister for Environment and Transport" in the coalition agreement to the appointment of a "Minister of Transport and the Environment" represents a change of emphasis or euphonic preference.

[65] *Partnership for Scotland: An Agreement for the First Scottish Parliament* (May 1999); the text was printed in several newspapers, e.g. *The Scotsman*, 15 May, 1999.

[66] Also, some of the individuals on the negotiating teams were known to have an interest in the topic, and the Liberal Democrats rely heavily on rural areas for their support.

[67] E.g. the promise to "tackle pollution".

Sustainable development has made several appearances in the new Parliament. The issue was the subject of a formal debate in February 2000,[68] leading to a resolution:

> That the Parliament places sustainable development at the core of its work and commends the Scottish Executive for its commitment to integrate the policies of sustainable development into all Government policies for the benefit it brings to the people of Scotland now and in the future.[69]

During that debate and before the Transport and Environment Committee, the Minister has referred to the ministerial team on sustainable development, including some external members, and to the forthcoming development of Scottish indicators for sustainable development,[70] but as yet there is little to show exactly how thoroughly the Parliament and Executive are going to give effect to their rhetoric.[71] The policy memorandum which accompanies all government bills has to include a statement on the impact of the bill on sustainable development,[72] but so far these have hardly provided a deep analysis of the potential implications of the proposed measures.

Environmental matters also feature strongly in the legislative programme announced by the Scottish Executive.[73] Legislation has been introduced to create National Parks in Scotland, and is proposed for a wide range of land reform measures, including the right of responsible access to land for recreation and passage. Both of these measures arise from thorough consultation exercises during the past few years.[74] A further bill is proposed for transport matters, introducing powers for road-user charging and workplace parking levies and reforming the regulatory framework for bus services. Fitting together such measures, largely exercisable by local authorities, with the powers in the hands of the Scottish Executive to provide a coherent transport strategy for Scotland in the context of the decisions taken at Westminster in relation to reserved

[68] 3rd February, 2000; Official Report vol.4 cols.781ff.

[69] Ibid., col.827.

[70] Building on the UK indicators discussed in *A Better Quality of Life: A strategy for sustainable development for the UK* (Cm.4345, 1999) chap.3.

[71] There is no statutory duty on the Parliament or Executive in this regard, unlike the National Assembly for Wales which is under a duty to prepare a scheme for promoting sustainable development (Government of Wales Act 1998, s.121); see the Assembly's consultation paper *A Sustainable Wales – Learning to Live Differently* issued in January 2000.

[72] Standing Orders (edition 1; 10 December, 1999) Rule 9.3, para.3(c).

[73] Scottish Parliament Official Report, vol.1 cols.403ff (16 June, 1999).

[74] Consultation documents include: *Developing Proposals for National Parks for Scotland* (SNH, 1998) and *National Parks for Scotland: A Consultation Paper* (SNH, 1998), leading to *National Parks for Scotland: Scottish Natural Heritage's Advice to Government* (SNH, 1999); *Land Reform Policy Group: Identifying the Solutions* (Scottish Office, 1998) and *Land Reform Policy Group: Recommendations for Action* (Scottish Office, 1999) leading to *Land Reform: Proposals for Legislation* (Scottish Executive, 1999).

matters (e.g. road fuel duty) will be a challenge. A private member's bill to prohibit hunting with dogs has also been introduced.[75]

Over time, the social and economic context in Scotland will inevitably lead to
different priorities and initiatives. Fishing, agriculture and forestry are more
significant parts of the Scottish economy than of the United Kingdom as a
whole, and a much greater proportion of Scotland is subject to conservation
designations.[76] In many areas across Scotland there is an urgent need to encourage economic development, rather than to grapple with the consequences of
excessive development pressure as in parts of the south-east of England. With a
different density and distribution of population, transport pressures and needs
are different, as are those affecting the water supply and sewerage systems.[77] As
one of many possible examples, the recent consultation papers on the future of
Sites of Special Scientific Interest (SSSIs)[78] reflected different approaches to the
issue on both sides of the border even before devolution took effect.

That last issue has also revealed what is likely to be a further cause of difference in the law throughout the United Kingdom. It is inevitable that with
different pressures on the time of Parliaments and administrators the enactment of new laws and the formal adoption of policies will fall out of step, even
where there is no substantive difference of approach. For England and Wales,
the measures giving effect to changes in the law on SSSIs have been included in
the Countryside and Rights of Way Bill introduced to the Westminster Parliament in March 2000, whereas the priority given to National Parks has meant
that as yet there are not even formal proposals for legislation in Scotland. After
years of consultations, regulations to give effect to the new regime for contaminated land have been introduced for England[79] but not yet in Scotland or Wales.
On the other hand, the *National Waste Strategy: Scotland*, produced by the
Scottish Environment Protection Agency, was published in December 1999,
before its English counterpart. The different administrative and legislative
timetables across the UK will lead to a fragmentation of the law and policy,
even where parallel measures are being adopted.

---

[75] Protection of Wild Mammals (Scotland) Bill.

[76] In Scotland 11.3% of the land is designated as Sites of Special Scientific Interest, as
opposed to 7.4% of England and Wales, and almost 20% of Scotland falls within Environmentally Sensitive Areas (figures from: The Scottish Office, *A Review of Natural
Heritage Designations in Scotland: A Discussion Paper* (1996)).

[77] Generally, in Scotland there is not the same difficulty in meeting the demands for
drinking water as in the drier and more densely populated south, but major investment
in waste water treatment is required to replace past reliance on direct disposal of sewage
to the sea.

[78] The Scottish paper places greater emphasis on community involvement and the sim-
plification of the designations system, whereas the English one emphasises issues of
protection and management; Scottish Office, *People and Nature: A New Approach to
SSSI Designation in Scotland* (1998); Department of Environment, Transport and the
Regions, *Sites of Special Scientific Interest: Better Protection and Management* (1998).
See also paper by Last above.

[79] Contaminated Land (England) Regulations 2000, SI 2000 No.227.

The extent to which separate policies are pursued in Scotland will be influenced not just by such "objective" factors but also by political considerations. Even when the same political party is in power at Westminster and Holyrood, policy differences may emerge. If the political make-up of the two administrations differ, then wider differences will almost certainly appear. Where coalition arrangements are required in Scotland as a result of the proportional voting system, environmental matters may be seen as one of the more readily negotiable areas, where visible concessions to the lesser party can be made without threatening the core of the leading party's policy agenda.

Moreover, in the early days especially, there may be a tendency to seek difference for difference's sake, and thus not to support measures adopted by UK authorities without what is perceived as adequate recognition of separate Scottish interests. The arguments over the rise in fuel duty costs, as mentioned above, provide one example of this. To the extent that such an approach generates carefully considered policy initiatives fully adapted to the Scottish context, this is to be welcomed. But there are two dangers. The first is that there will be uncertainty and delays in carrying forward initiatives which commenced before devolution took effect. Many of the issues covered in the UK Government's policy statement *A Better Quality of Life – Strategy for Sustainable Development in the UK*[80] are ones which are now divided between the Scottish and UK authorities. The development of purely Scottish responses to these issues would take a major effort, and if existing measures are to be rejected or reconsidered simply because of their non-Scottish origin, then progress will be greatly slowed.

The second danger is that the emphasis on "Scottishness" may get in the way of the integrated approach needed to deal with so many environmental challenges. Few issues can be addressed in Scotland in isolation, and since the powers required to give effect to a meaningful environmental policy are divided between Holyrood and Westminster, it is essential that the authorities operate together. Achieving this will require sensitivity on all sides, and goodwill to ensure that sound working arrangements can be developed.

Conclusion

The devolution of power to the new Scottish authorities clearly opens the possibility of greater divergence on many issues.[81] The simple fact of administrative separation, meeting the needs of Parliaments which work in different ways and on different time-tables, will by itself tend to increase the extent to which Scots law is different in form and style, regardless of content. At the very least, legislation can be conceived from a Scottish perspective from the beginning, as opposed to "putting a kilt" on a model created to fit the English system.

---

[80] Cm. 4345 (1999).

[81] See generally, E. McDowell & J. McCormick (eds.), *Environment Scotland: Prospects for Sustainability* (1999, Aldershot, Ashgate).

The scope for wholly new initiatives in Scotland is limited by the extent to which modern environmental law emerges from international or European Community agreements. The degree to which Scottish environmental law develops as a distinctive system is thus likely to be more the result of the gradual accumulation of minor differences in emphasis or form, than the consequence of fundamental policy divergence from the rest of the United Kingdom. Nonetheless, there will be increasing differences, and the new beginning offered by devolution, consciously breaking away from many aspects of how business is done at Westminster, creates the opportunity for environmental issues to play a central role in all aspects of the government of Scotland.[82] On the other hand, environmental matters could be relegated to a marginal role, sacrificed to the needs of economic development and used as examples of the extent to which the Scottish authorities are not completely independent but forced to comply with commitments made by others outside Scotland. As with all aspects of the devolution scheme, much will depend on whether the determination to make the new arrangements work is strong enough to overcome the inevitable areas of dispute or political strain between all of those involved in, or subject to, the new settlement for governing Scotland.

---

[82] See, for example, *Scotland the Sustainable?: 10 Action Points for the Scottish Parliament*, a paper produced by the Secretary of State for Scotland's Advisory Group on Sustainable Development in March 1999.